D0705370

DATE DUE

JUST GENES

JUST GENES

The Ethics of
Genetic Technologies

Carol Isaacson Barash

 Westport, Connecticut
London

Library of Congress Cataloging-in-Publication Data

Barash, Carol Isaacson.
   Just genes : the ethics of genetic technologies / Carol Isaacson Barash.
      p. cm.
   Includes bibliographical references and index.
   ISBN 978-0-313-34900-3 (alk. paper)
   1. Genetic engineering—Moral and ethical aspects. I. Title.
   QH438.7B37 2008
   174′.957—dc22   2007035291

British Library Cataloguing in Publication Data is available.

Library of Congress Catalog Card Number: 2007035291
ISBN: 978-0-313-34900-3

First published in 2008

Praeger Publishers, 88 Post Road West, Westport, CT 06881
An imprint of Greenwood Publishing Group, Inc.
www.praeger.com

Printed in the United States of America

∞

The paper used in this book complies with the
Permanent Paper Standard issued by the National
Information Standards Organization (Z39.48–1984).

10 9 8 7 6 5 4 3 2 1

Contents

Preface

Advances in genomics are spawning revolutionary technologies that promise to dramatically improve medicine and agriculture, as we've known them. Excitement about the capabilities of these technologies runs deep, as expected benefits are predicted to be vast. Whether benefits will be available to all people is arguably a disturbing uncertainty, for rarely are innovations distributed equally. Those at the lower rungs of socioeconomic and educational status are often the last to gain access, if they do at all. Acknowledging the likely disparity in distributed benefits is, arguably, the starting point for informed and ethical policy debate about appropriate use and access. Ultimately, each of us has an important participatory role in ensuring the just use of these advances.

Many have called attention to the proactive approach the Human Genome Project, and its subsequent research initiatives, has undertaken—as evidenced by the establishment of the ELSI agency (Ethics, Legal and Social Implications Division, of the National Institutes of Health)—to identify and recommend proactive measures to prevent the inappropriate use of these advances. While ELSI's work has led to public policy at national and state levels, as well as to clinical care guidelines recommended by professional organizations, the ways in which genomic advances will be adopted will continue to evolve for some time. Foreseen, and unforeseen, ethical challenges will inevitably emerge in the course of this process. The issue of fair access to the expected breakthroughs will remain potent and contentious.

Ensuring just distribution of genomics benefits, access, and appropriate use necessitates an informed citizenry. Such a citizenry understands not only the clinical science but also the ethical dilemmas posed in using diagnostics, therapeutics, disease risk management strategies, or new modified foods. This book was written with the aim of promoting such a citizenry. It offers the reader the opportunity to learn how to better identify and analyze ethical issues arising in several areas in which genetic advances are already affecting health care and agriculture.

My wish is that the reader achieve a deeper appreciation for human's long-standing interest in hereditability coupled with not only an understanding of the complexities of ethical issues that arise in adopting these new genetic technologies, but also the importance of not reifying genes as the final arbiter of just usage.

The idea for this book originated with the desire to provide a compilation of genetics-ethics topic areas that would educate the reader about basic principles of moral theory and applied ethics, while also offering exercises that permit experiential learning when grappling with hypothetical ethical dilemmas of the sort discussed in each chapter. My goal was to offer in a single book, methods of ethical analysis, an introduction to some of the many ethical challenges we—individually and collectively—face in adopting genomic advances, and the basis for students to apply those methods in each topical chapter. It is my hope that my framing the current ethical issues through an historical lens will be instructive and propel us as a society to ensure the just distribution of the benefits anticipated.

Acknowledgments

The final manuscript would not be possible without Dr. Priscilla Short's dogged commitment to every detail, her help, and her encouragement. Her conceptual framework, scientific input, and thoughts were pivotal to the final chapter. I also thank the following people for their extremely helpful comments: Dr. Daniel Herwitz, Dr. Hilary Davis, Dr. Vicky Whittemore, Dr. Ann Kindfield, Dr. Doreen Blanc, and my editors at Praeger, in particular Suzanne Staszak-Silva, without whose exceptional input and hard work this book would not be possible.

Introduction: Are We Ever Talking Just about Genes?

Thanks to media reporting, you don't really need to understand what a gene is, or does, to know that a lot of other people do. Advances in genetics and molecular biology have spawned new subfields (such as genomics, proteomics, glycomics nutrigenomics, toxicogenomics, and so on) and related fields (computational biology). Hundreds of research institutions around the world are identifying and analyzing various aspects of genes, including single nucleotide polymorphisms (SNPs), haplotypes, functionality, by-products and expressivity under specific environmental conditions, dynamic relationships with a broad array of biological processes, and more. Research findings and technological applications have spawned the creation of hundreds, if not thousands, of companies that are developing commercial products and services designed to offer improvements in medicine, agriculture, reproduction, forensics, ancestry, nutrition, skin care, athletic performance, and other domains. These advances coupled with intersecting fields (for example, nanotechnology, e-health, robotics, electronic bio-medicine, and photonic bio-medicine) are expected to revolutionize health and agriculture in ways that are only just now beginning to be imaginable. More and more people are using genetics in their professional and personal lives, particularly as the availability of genetic testing continues to expand with, for example, cord blood and stem cell applications concretizing the future. The term "genetics" has arguably morphed from a rarefied term of the scientific elite to a common household word.

Begun in 1990, the Human Genome Project (HGP)—the multinational quest to identify, map, and sequence every single gene contained within the double helix of humanity's DNA (Deoxyribonucleic acid)—has laid the foundation for much of what we are now learning about the genetic influences of various diseases and traits. From the 2001 completion of the draft of the Human Genome, we now know that the human genome is made up of roughly 30,000 genes and that these genes influence, albeit to varying degrees, the development and expression of various human characteristics. The project has been characterized as the most ambitious science project since the U.S. Apollo program to land a man on the moon and has been dubbed the "holy grail of biology." It set in motion a huge promise to revolutionize medicine through therapeutics designed to mitigate suffering incurred from genetic disease, enhance disease prevention capabilities, and personalize health interventions and avoidance strategies based on identifying predispositions and hereditable susceptibilities. It also enabled a wide range of explorations and technological developments involving the genetic basis of numerous human characteristics not directly relevant to clinical medicine, including but not limited to ancestry, breeding, behavior, personality, language—essentially the mysteries of all plant and animal life. The personal and societal implications of this explosive growth in our knowledge about genes extend far beyond the biology.

The days of just predicting impact is nearly gone, for genetics, if not also genomics, is affecting us now. A broad range of commercial applications already exists in the marketplace. The contexts in which genetics appear keeps expanding, from a few pharmacogenetic tests and an array of DNA based diagnostics, to the future promise of safer and more effective therapeutics, to more sophisticated forensics, identity, and ancestry testing, to companies touting the use of DNA in their advertising,[1] perfume,[2] underwear,[3] or even personalized anti-aging skin care products.[4] The extent of future innovations seems limitless. The impact of genetics on daily life is already dizzyingly broad in scope, even though so many unknowns remain. There is little reason to believe that the impact will lessen as more discoveries and applications the fields advance.

Beyond this vast exposure, though, is the uncertainty of exactly what the use of genetics will mean for each of us individually and in relation to others on a familial, societal, or global level. Knowing your genotype, and therefore something of potent relevance to you, may well influence your view of your life opportunities. Learning a little, or even a lot,

about your unique genotype will likely affect you in ways you can anticipate, but also in ways you can't. For example, knowing how you respond
to certain environmental toxins or which diseases you are at high risk
for may well impact your view of your ability to achieve your educational or athletic goals.

Have you ever pondered whether you would or would not want to
learn your **genotype** and under in what circumstances you might want
or not want to know? Can you imagine what knowing your genetic
makeup (or genotype) might mean to you? Have you ever wondered
whom in your family you are most like, or whether you might have
inherited a particular trait or talent from either of your parents that they
hadn't developed? Or whether you harbor mutations that increase your
risk for common diseases, like cardiovascular disease or cancer? Or
wondered how much of your food is genetically modified (GM) food?

As the fund of human genetic knowledge grows, it is harder to ignore
the pervasiveness of genetics. With so much of what we observe and
experience in life potentially tied to genetic influences, the impact of
genetics on various facets of our life continues to deepen. If not from
reading the paper or a magazine, listening to the radio, or watching
television reports about new genetic research findings, we may be cogni-
zant of ads selling DNA brands or insinuating that we got the "genes"
for those jeans. We think and talk as if genes have absolute power over
us. We might glibly chide ourselves for having the gene for exercise
phobia or a shopping addiction. Even if we understand enough to know
that **phenotypic expression** is far more complicated than just a gene firing off a protein to produce a trait (there really is no "exercise gene" or
"shopaholic gene," as if a single gene causes behavior), plenty of us still
gravitate to talking about genes as if they were mini pistols firing out
specific phenomenon or dispatchers directing our lives. Because genes
exert significant influences, there's a common tendency to regard these
infinitesimally tiny, but powerful, parts of our being as encapsulating
our innermost essence. Such talk about genes encoding for this or that
trait lends to an aura of inevitability as though our capacities are located
in the executing centers of those "masterful molecules" the genes.
Whether we mean what we say or not, it's a frequent enough usage that
it reinforces the deceptive perception that genes have absolute power
and importance in shaping our lives. To believe that our genes define
our essence as humans is misguided. However basic and integral to life
from inception to death, our genetic endowment is but one of several
threads that together weave the fabric we call life. Clearly, reducing
human essence, experience, or purpose to a string of nucleotide bases

misses the meaning we give and take from our existences on earth and the richness of life in its entirety. Any view of genetics that bestows the human genome with a deterministic power is not the whole story, as will be discussed.

Perhaps, the paradox in unearthing genetic secrets in personhood is that individual humans differ comparatively little from one another in their genetic makeup, just 0.1 percent. Furthermore, the human genome is remarkably similar to that of other species. The genotypic difference between humans and chimpanzees, for example, is just 1.6 percent. A deeper understanding that our shared humanity (genetic similarity) is far greater than our difference could serve as a force to unite people in peace, thus ending perpetual hatred and violence toward one another. But there is little, if any, indication that it will.

Once we start talking about how new genetic innovations will affect us, or how they ought to be used, we are immediately and inextricably involved in a discussion that is more normative and political than it is scientific. This is so because—whether or not we are aware—such discussions are infused with a host of prescriptive notions of what is right and wrong, which scientific applications ought to be pursued and which directions foreclosed, beliefs about where the line between normality and abnormality is drawn, attitudes about what a disability is or isn't, or opinions about who is or isn't deserving of which benefits and protections from harms. Furthermore, fundamental questions arise about whether we ought to undertake activities just because we can. These concerns sharpen debate about the role genetics should have in defining who we are, how we ought to improve the earth, and even what type of world we want future generations to inherit. A goal of ethicists and policy analysts is to identify and evaluate those latent assumptions in efforts to clarify policy questions and frame the debate for the best possible deliberation about policy setting.

One such threshold question is what in fact is disease and what isn't. Is a genetic risk, like an inherited susceptibility to cancer later in life, a disease in its own right? Or a disability? Even if asymptomatic? Or only when symptomatology sets in? How much symptomatology is needed before the state of health is appropriately referred to as illness and no longer qualified health. Ought presymptomatic disease and traditional disabilities like deafness be similarly protected under the law? Should we strive to eliminate as much genetically induced disease as possible? Individuals and their families can suffer enormously when a genetic disease strikes a family member. If not for the sake of the individuals directly affected, should we eliminate disease because it is an avoidable drain on scarce social resources? Should we eliminate only lethal genetic diseases? And if so, which diseases ought to be deemed "lethal" and on what basis? What if today's "lethal" diseases become tomorrow's chronic diseases, thanks to improved treatments? Should that likelihood change our view?[5] Furthermore, given the expectation that genomic advances will lead to new therapies capable of fully treating or substantially mitigating various genetic diseases, under what conditions, if any, might it be appropriate to avoid the risk, or the certainty, of genetic disease? If we decide that we should seek such ends, how ought we to accomplish them? Should we endorse prenatal testing and pre-implantation diagnosis? Should cost be a factor in determining which technologies to use? Who should pay: insurers or patients? Should we institute programs that encourage, if not require, pairing between genetically compatible people to avoid the risk of disease all together and the expensive technologies to eliminate it? Such an idea is not entirely farfetched. And all of these questions are just a small fraction of the myriad ethical concerns that arise.

Consider, for example, that for several decades states have *required* couples planning marriages to undergo testing for syphilis and rubella because future pregnancies could be negatively impacted by intrauterine transmission, and states want to protect possible offspring from adversity. Premarital Tay-Sachs screening (of Jewish and recently French Canadian couples) has been encouraged to permit couples to learn whether they are both carriers (in which case they carry a 25 percent risk of conceiving an affected child with each pregnancy) to permit them to make more informed reproductive decisions. A community-based program in existence for years in one orthodox Jewish group encourages young adults around the age of 15 or so to undergo screening for a number of genetic conditions for the purpose of encouraging dating only among those whose coupling poses no genetic risk. With state deficits increasing and escalating costs of caring for and accommodating disabilities, it is not preposterous to imagine a state legislature mandating certain types of premarital, preconception, or prenatal tests as a strategy to lessen societal financial burdens. Such was preliminarily attempted in Colorado in 1993 when the legislature considered a bill to mandate widespread **Fragile X** testing on the basis of studies (although flawed) indicating that each affected child costs the state well over a million dollars during their lifetime. Genetics, in fact, has been invoked, rightly or wrongly, to solve social problems, such as paternity testing to identify dead-beat dads and make them pay up as a case in point. This and similar concerns about the extent to which the state or communities ought to curtail personal liberties through mandated genetic testing for the sake of larger societal benefit will be taken up more fully in subsequent chapters.

Similarly, controlling for the prevention of genetic disease raises interesting larger questions of not only which diseases, but which *traits*, ought to be prevented. Deafness, for example, is generally considered a trait, *not a disease*, but nonetheless is viewed by many (though not all) as a liability. There are many people who if given a choice would wish to avoid giving birth to a deaf child. Some view deafness as not just a hardship for affected individuals but also as a possible burden on families and a drain on scarce societal resources. What many people perhaps do not know, though, is that within the deaf and blind communities, deafness and blindness are not necessarily viewed as profound liabilities, and therefore many deaf and blind individuals prefer to maintain their community culture and genetic legacy by birthing children who are also deaf, or blind, thereby ensuring a common heritage from generation to generation and an ensuing emotional closeness. Individuals with **achondroplasia** have articulated similar concerns.

Furthermore, controlling for prevention of genetic disease through the use of prenatal and **pre-implantation diagnosis** illustrates what some consider an ethically suspect strategy, namely phenotypic prevention by genotypic prevention. That is, identifying the genetic risk preconception or the genetic diagnosis prenatally permits the avoidance of birthing affected children. Such a preventive strategy relinquishes any societal imperative to development treatment, because if affected children aren't born, treatments need not be developed for those who are affected. Affected individuals, then, stand no hope that their disorders could be better treated or cured.

Genetic enhancement raises converse questions. Should we optimize our genetic lot through enhancement strategies, like genetically modifying the food supply or the human germ-line to prevent disease? When, if ever, is enhancement ethically acceptable? Most parents work hard *to enhance* their children's likelihood of success through a wide range of tactics: extracurricular lessons and training, fashion and fitness to ensure good looks, tutoring and prep work to guarantee acceptance at a good school. Athletes prime their ability to perform by modifying their diet, using computer training regimens, and sometimes even taking steroids, drugs, or other enhancers.

Similarly, we generally act upon the belief that avoiding adversity is a good thing and thus get vaccines to prevent lethal disease, take prophylactic medications to prevent potentially lethal infections from exposures, wear seat belts to guard against traffic fatalities, and so on. Whether genetically based enhancements are different in kind or degree from other enhancements is but one of many controversies surrounding the debate about what genetic enhancement is and whether it should ever be permitted. Additionally, who would have access to genetic enhancement raises the issues of the distributive justice. Most likely, only those who can afford enhancements would be able to improve their genetic endowment, although the potential for intentional or unintentional coercion ought not be minimized, as will be discussed in subsequent chapters.

Most geneticists believe that genetics ought to serve human ends. But, we must first ask whether this belief itself is ethically justifiable. Why should genetics serve human ends? Which goals ought to be pursued and why? Clearly, moral contention arises in determining such ends, as well as deciding whether any of them ought to be achieved, much less how. Subsumed within the debate are competing values, though these values are not always obvious or even acknowledged. Primary to assessing ethical considerations that may arise is identifying specifically what values underlie the debate and that are at stake in considering the adoption of genetic innovations.

The central aim of this book is to explore what values, ethical principles, and assumptions underlie the many and varied policy questions that arise from determining how best to integrate genetic innovations (new genetic knowledge and technologies) into practice. In so doing, a primary aim of this text is to challenge the reader to identify, clarify, and analyze the normative and political underpinnings of debate about the use of genetic innovations. How important are our genetic underpinnings in the greater mix of life? Will the "genetics revolution" truly revolutionize life as we know it? More important, do we want it to? What kind of say do we, as ordinary citizens, want in determining which innovations ought to be used and how?

Several books have addressed ethical issues associated with the new genetics. Few if any have attempted to directly increase the public's understanding while also improving on the related ability to critically evaluate claims and assess issues. This book hopes to sharpen the reader's critical and analytic faculties to better evaluate ongoing issues, whether they are media reports or personal decisions about whether to participate in a form of genetic testing, purchase a genetically modified food, or engage in public debate and policy making. With each chapter comes the opportunity to learn not only what ethical issues pertain to this new era but also how such issues potentially affect you, your family, your local community, and the global community. Additionally, readers can improve their reasoning by working through the exercises at the end of the book. If you find that this is a book of questions not answers, it is in fact just that. That's because the goal of this book is to provide you with a way to think about what the answers are. It is in no way a claim to have convinced you what's right or wrong.

The ethical issues discussed in this book are by no means the only ones. Nonetheless, at the end of this book, you should better understand all the hoopla about genes. Ideally, you will then be better equipped, both on a personal and civil level, to critically assess media coverage of genetic advances, know what some of the key ethical issues are surrounding the use of this new knowledge and technology, identify underlying assumptions driving your opinions, analyze the merits of your position, and participate in policy debate about the use and misuse of genetic technology and genetic information. It may well be our duty as citizens to educate ourselves and actively participate in the determination the choices that lie ahead regarding genetic applications.

Our genes surely are not fair. It's hardly fair for some to have severely deleterious genes and others not. But genes are not our destiny, unless, of course, we allow them to be. The value we place on human genetic variation and expression will depend not only on **somatic** and **germ-line** replications but also on notions of fairness and justice—on an interpersonal, communal, societal, and global level. The time to get it right is *now*. Genes, after all, are just genes—nothing more, nothing less. And yet we hardly are ever talking just about genes.

Notes

1. Novinky DNA Communications Group, http://www.dna-advertising.cz; DNA Advertising, art.youngliving.com.

2. DNA Bijan Eau de Parfum Spray, www.1-click-perfumes.com/women/dna-w.htm DNA.

3. DNA Products, LLC, www.dnaproductsonline.com.

4. Lab21, Inc., http://www.lab21.com/new/dna-faq.htm, http://www.lab21.com/web/about-faq.php.

5. For example, one can anticipate that in the future cystic fibrosis (CF), which currently is likely to confer a shortened life span for those affected, will be consistent with a normal life span equivalent to other chronic diseases. Given this likelihood, is it justifiable to choose not to give birth to a fetus prenatally diagnosed with CF?

CHAPTER 1

Justice and Genes: A Brief Historical Overview of the Relevance of Ethics to Human Genetics

Ethics is the study of moral principles. Genetics is the study of biologically inherited traits. That the two would be linked in any way might seem fantastical at best, yet they have been intertwined for centuries. To be sure, human curiosity spurred increased knowledge of heredity knowledge for its own sake. But, concomitant to this was the drive to use this knowledge to better society.

The Science of Human Heredity

People have pondered heredity for roughly 6,000 years. The earliest evidence is found in Babylonian engravings of pedigrees documenting the transmission of specific characteristics of horses' manes. The ancient Greeks continued to build on that early knowledge. Aristotle and Hippocrates, in particular, theorized that important human characteristics were determined by semen, which used menstrual blood as a cultural medium and the uterus as an incubator to produce the human body. Males, not females, were thought to determine patterns of inheritance, with females contributing in a negligible way. All characteristics, or phenotypes, were believed to be determined by male contribution. In other words, male contribution dominated the hereditability of all traits. Bald men would produce bald sons, for example. Or, if women were bald, they had inherited baldness from their fathers. This belief, that only men influence heredity, prevailed for nearly two thousand years. It wasn't until the Scientific Revolution in the late 1700s that scientists recognized the existence and interactivity of sperm and ova, thereby offering an explanation for how a female could transmit characteristics to her offspring.

For many, it's hard to think about genetics without harkening back to Gregor Mendel, the Austrian monk credited with the first explanation of the laws of heredity in 1865, from his exhaustive work with peas. His findings were lost but then rediscovered and laid out in Morgan and colleagues' *The Mechanism of Mendelian Heredity*, published in 1915.[1] Mendel postulated that "units contributed by two parents separate in the germ cells of offspring without having had influence on each other."[2]

From his experiments crossing yellow peas with green peas, he observed that half of the seedlings were yellow and the other half green. From patterns in observed phenotypes (organismic traits) from generation to generation, Mendel reasoned that the patterns of transmissibility conformed to certain rules. His **monohybrid** and **dihybrid** cross studies demonstrated that genes, not the traits themselves, were transmitted. These cross studies further illustrated the principles that govern the inheritance of both a single (monohybrid) trait and two (dihybrid) different traits.

The science of heredity came into its own in the late 1800s with the demonstration that distinct traits such as polydactyl (extra digits) or albinism (lack of pigmentation) were inherited differently. In the 1870s, scientists observed that the cell nucleus of male and female reproductive cells undergo fusion in the process of fertilization. This finding led to the identification of chromosomes and their "splitting" behavior (meiosis). Morgan demonstrated that Mendel's "unit of inheritance" was carried in chromosomes in the sperm and egg in one half complement (1N), which when fused reconstitutes the normal 2N of chromosomes.[3] Simultaneously, it was discovered that chromosomes contained DNA (Deoxyribonucleic acid) and proteins. Initially, it was thought that other proteins were the transforming material. In the 1920s, this transforming material was discovered to be DNA contained in the chromosomes. Notably, it was around this time that the word *gene* was given a permanent place in the dictionary by Wilhelm Johannsen, a Danish biologist. The word *gene* comes from Greek meaning "to give birth to." A human gene was first assigned to a chromosome in 1968, roughly a half-century later.

Scientists did not, however, wait for these twentieth-century clarifications to apply genetic concepts to the goal of improving humankind. Crossbreeding of crops and animals, for example, had been done for hundreds of years, though largely by accident or intent. *Controlled* breeding, that is, the possibility that one could cross organisms (engineer the manipulation of genes) to produce certain traits, was a possibility implied by Mendel's theory of dominance and recessive inheritance. Such newfound ability raised questions about if, or when, manipulating breeding was acceptable, and where the line between acceptability and unacceptability should be drawn. In other words, even at the field's earliest beginnings, there was great zeal about the social utility of genetics.

Applying Principles of Heredity

Thousands of years ago, farmers selectively crossbred for domesticated animals and plants to produce food that possessed specific desired traits[4] and, in the case of animals, breeds that displayed certain desirable characteristics and behaviors. Similarly, people have sought to improve their own lot, as well as that of their future offspring, in various ways. Marriage and mating decisions were made by families to produce desirable phenotypes, such as good looks, intelligence, longevity, and other characteristics deemed advantageous, while avoiding undesirable phenotypes such as serious disabilities, lethal disease, ugliness, and the like. But the role of genes was not well understood, so the success of breeding techniques depended on luck and other uncontrollable influences, rather than on engineering knowledge.[5]

The enticing idea that bad traits could be eliminated by preventing the breeding of bad strains thus emerged as feasible, if not also appropriate. Predictive hereditability could, for example, enable the breeding of a specific crop for its increased yield, flavor, and resistance to disease. Similarly, a logical solution to demonstrable failures of various social reform programs was to transform the perceived fixed cause, namely, genes. Excitement about applying Mendel's laws of predictive hereditability to the task of eliminating chronic societal ills arose. In other words, prevent phenotype by avoiding genotype.

Sir Francis Galton coined the term "eugenics." Eugenics literally means "well born" and refers to the desired improvement of humans' genetic endowment over time. The goal of eugenics was to improve humankind by harnessing knowledge of human heredity and applying it to social programs.

In 1893, Galton wrote:

> What nature does blindly, slowly and ruthlessly, man may do providently, quickly and kindly. As it lies within his power, so it becomes his *duty*

(emphasis added) to work in that direction; just as it is his duty to succor neighbors who suffer misfortune. The improvement of our stock seems to me one of the highest objects that we can reasonably attempt.[6]

It was not long after that the Eugenics Society of London and its U.S. analogue, the Eugenics Record Office at Cold Spring Harbor (Long Island, New York), were established to achieve the goals of eugenics. Those goals were both positive (the expansion and enhancement of desirable traits) and negative (the elimination of genetic disease and undesirable traits).

Charles Davenport, the director of the U.S. Eugenics Record Office, zealously embarked on research intended to provide the scientific basis, and therefore moral defensibility, for operationalizing eugenics' goals. He investigated the hereditability of numerous human "traits," specifically criminality, poverty, feeblemindedness, moral degeneracy, and genetic diseases. Observing that these traits seemed to run in families, Davenport constructed pedigrees based on Mendelian transmission principles, believing that these traits were purely genetic and not amenable to mitigating environmental influences. For example, if an individual was born into poverty, Davenport believed he had proved why this individual would remain poor throughout life. Similarly, if one's parent were a criminal or prone to moral degeneracy, this individual had a particularly high probability of committing a crime and/or leading a life of debauchery. At the time, scientists did not offer countervailing evidence, and so gross overestimates of the role of genetics in behavioral traits were uncritically accepted as fact.

In the early twentieth century, U.S. eugenicists strongly influenced social policy in the name of negative eugenics, or the elimination of undesirable traits. Their "scientific testimony" led Congress to pass discriminatory immigration laws and states to enact involuntary sterilization laws. The first such sterilization law was enacted in Indiana in 1907. It forbad feebleminded persons from ever reproducing by requiring that they be sterilized, based on the "certainty" that, if they reproduced, they would produce only feebleminded offspring. Although subsequently this, and similar laws in other states, were declared unconstitutional, the U.S. Supreme Court upheld the view in its 1927 *Buck v. Bell* decision, which involved the institutionalization and sterilization of a patient, Carrie Buck, who had been declared feebleminded, as were her mother and her daughter. Although never medically examined, an "expert" in the case found the women to be "genetically defective." Furthermore, even though the daughter's teachers found her to be bright, the so-called expert nonetheless prevailed in arguing her mental deficiency. Justice Oliver Wendell Holmes famously declared that

> we have seen more than once that the public welfare may call upon the best citizens for their lives. It would be strange if it could not call upon those who already sap the strength of the State for these lesser sacrifices, often not felt to be such by those concerned, in order to prevent our being swamped with incompetence. It is better for the entire world, if instead of waiting to execute degenerate offspring for crime, or to let them starve for their imbecility, society can prevent those who are manifestly unfit from continuing their kind. The principle that sustains compulsory vaccination is broad enough to cover cutting the Fallopian tubes. Three generations of imbeciles are enough.[7]

Not only was the case factually incorrect, but the decision is dubious on moral grounds. The genetic suppositions underlying the "facts" put forth were largely incorrect. Mental retardation is not a single entity, nor is it a genetic disease caused by a dominant single gene. We now know that there is a wide array of mental deficiencies, some of which are due to a major gene with differing modes of inheritance, some due to chance polygenic (more than one gene) combinations, others due to chromosomal aberrations, and yet others due to environmental causes in development and birth, such as oxygen deprivation during birth (cerebral palsy) or the presence of in utero viral infections, including toxoplasmosis, hepatitis B, herpes zoster (which causes chicken pox), syphilis, rubella (which causes German measles), cytomegalovirus, and herpes simplex. More important, even *if* the "mentally deficient" did not reproduce, the frequency of mental deficiency in the population would decrease only slightly, if at all (with major emphasis on the word "if").

Negative eugenics, and its legacy of forced sterilizations of persons deemed "unfit" or unlikely to have socially adequate offspring, garnered such strong support that today 19 states retain statues permitting eugenic sterilization of institutionalized retarded persons. Although such laws are seldom (if ever) applied, their continued existence remains a chilling reminder of how misapplied science, even if well intended, can promote grotesquely immoral consequences.

Positive eugenics, or the expansion and enhancement of desirable characteristics, represents the converse tactic for promoting human betterment. In the early 1900s, "family fittest" contests were held throughout the United States in search of families who exemplified superiority. Those with obvious talent, superior health, moral character, genius, and/or good looks were deemed to be not only of high "civic worth," but *obligated for the sake of all* to increase their propagation. Positive eugenics was thought achievable through educational programs that influenced public opinion, and particularly engaged wealthy individuals with poor but promising individuals. Galton even recommended that exceptionally worthy couples be provided convenient low-cost housing as an incentive to reproduce.[8]

Today, we know that these "desired" outcomes of positive eugenics could no more be engineered than those of negative eugenics. Increased procreation of the "fittest" cannot guarantee the absence of genetic disease, imbecility, or even moral character, because the complexity with which genes interact with other genes and environmental influences defies engineering phenotypes or outcomes.

Genetics as a Molecular Science

The foundation of genetics as a molecular science dates back to 1869 with the work of Friedrich Miescher who first succeeded in extracting nuclein.[9] He discovered a "weak acid" that was abundant in the nuclei of white blood cells; this chemical is now known as DNA. The observation that the cell nucleuses of male and female reproductive cells undergo fusion in the process of fertilization led to the identification of chromosomes and meiosis. Thus, the functional relationship between DNA and chromosomes was beginning to be better understood as was that of inherited single-gene diseases.

Genetic knowledge progressed steadily, but not rapidly. In the 1940s, Beadle and Tatum taught us that the function of genes was to produce proteins and enzymes that in turn propel chemical reactions that are the core of sustaining life. It was not until 1953 that Watson and Crick, using the x-ray crystallography findings of Franklin, learned more about what a gene was made of, what it looked like, and how it worked. The identification of genes as a double helix of nucleic acid was revolutionary, because it depicted how the sequences of bases specified the sequence of amino acids that the cell is to follow when creating a protein. Understanding how DNA replicates itself at the time of mitosis and meiosis, how transcription of DNA into RNA could then be secreted into the cellular cytoplasm from the nucleus, and how it could then be transcribed into proteins, opened the door to the age of recombinant DNA technology. That ability to manipulate genetic material ushered in the dawn of more precise genetic engineering.

Initially, this knowledge permitted the manipulation of genes whose protein products were already known, such as insulin. Working backward from the amino acid sequence of a protein to the messenger RNA, back to complementary DNA (cDNA), and finally to the gene for human insulin was relatively simple. Knowing the exact location of that gene in the full complement of human chromosomes, however, was still largely unknown. It then became the goal of scientists to have the full sequence of the human genome known at the individual base level. It would be the book of life spelled out at that the basic molecular level. From that point, scientists could determine what every gene comprising a human being was and where it was located. While scientists can extract DNA, using a medium that allows the extracted DNA to enter cells of another organism and effectively target where the DNA should be inserted in another cell, they cannot completely control the integration of the DNA. Thus, the controversy exists regarding the safety and benefits of these techniques, particularly in regards to germ cells that are passed on to subsequent generations. Only somatic cells, which are not inherited, are used to ensure that insertions do not contribute to the next generation. There are many possible outcomes of an insertion, including but not limited to (1) the DNA can fail to be integrated into the target location (or cells) thereby becoming **episomes**, (2) the DNA can become integrated into the cytoplasm in which case it may or may not be duplicated during cell division, or (3) the DNA may be destroyed and disappear all together. Genetic engineering is not the only way in which new DNA is integrated into host organisms. Common viruses are incorporated into the DNA of our cells all the time, such as each time we contract a viral illness. Nonetheless, recombinant DNA techniques have been successfully used to produce valuable therapies, particularly hormone replacement therapies, including insulin for diabetics, growth hormones for pituitary dwarfs, and erythropoietin for people suffering from severe anemia.

Molecular genetics came into its own at a time when technology was embraced wholeheartedly and with little reservation. Slogans like DuPont's "Better Things for Better Living … Through Chemistry" and Disney's "Our Friend the Atom" heralded the era of technology. With seemingly boundless support for technology's capabilities, society came to expect technology to enhance daily living and, especially, to solve societal ailments. Belief in the promise of *molecular genetic* technologies fit right in.

Genetic Testing and Screening

The term *genetic screening* refers to widespread testing of persons having a specific genotype without attention to or knowing their family history. Screening generally serves the two goals of public health, namely, (1) to detect a severe genetic condition before debilitation develops in efforts to mitigate disease by targeted therapeutics, and (2) to identify carriers, also referred to as heterozygotes (healthy, unaffected individuals who carry alleles for the genetic disease) so they can be informed of that risk for their offspring and may make fully informed reproductive decisions. The promise of technology to solve societal became actualized when scientists proposed a dietary strategy to cure the pernicious genetic condition, phenylketonuria (PKU). Enthusiastic support among geneticists spread to the public. Newborn testing, and therefore early detection and treatment (the diet), marked PKU as the first genetic disease with a cure.

The ability to prevent the severe mental retardation that comes from untreated PKU ushered in the rationale behind newborn screening, and such screening marked the first established widespread genetic screening program. PKU is an autosomal recessive disorder first described in the mid-1930s, 30 years before screening began. Two children who appeared completely normal at birth became severely mentally retarded and gave off a foul mousy odor. In 1947, this newly discovered disease associated with mental retardation was determined to result from a non-existent or minimally active enzyme that converts phenylalanine to tyrosine. This enzymatic dysfunction results in the accumulation of excess phenylalanine in the blood plasma, which, if the process is not reversed, results in disease. Scientists theorized that if the right amount of phenylalanine (an amino acid found in meat, fish, dairy products, and the artificial sweetener aspartame) was removed from the diet, then the disease would not develop. In 1963, the simple heel-stick blood test (known as the Guthrie test) was devised to detect elevated phenylalanine in the blood of newborns. Shortly thereafter all newborns were being tested for PKU at birth, and those affected were put on the special low-protein diet. In other words, this represented the first time a genetic disease had ever been conquered and, as such, it marked a tremendous advance in medical genetics.

While the vast majority sees the PKU story as a great American medical genetics success story, naysayers have argued that PKU is a qualified success and, in truth, that it is a greater political success than a medical one. They claim that diet does not prevent total morbidity for all those affected, treatments are not always covered by insurers, and going off the diet for pregnancy is problematic. Nonetheless, the dietary intervention created statewide interest in prevention, which in turn spurred increases in funding genetic research focused on screening.[10]

The PKU story represents yet another instance of applying genetics to solve a societal, albeit medical, problem, namely the incidence of mental retardation. PKU testing is now mandated in all 50 states as one of many newborn health screens. Yet the implementation of the screen, and its indicated treatment, is not without ethical, social policy, and legal controversy. Does the state have a right to require individuals (in this case parents for their infants) to undergo genetic testing? While newborn screening for PKU is mandated in every state, some states require that parents provide informed consent (and thus the possibility of informed refusal). Other states do not require consent; they just test all babies. Failure to obtain consent, particularly in light of state ownership of specimens (theoretically at least available for additional testing), raises a conundrum. Moreover, even though informed consent is a hallmark of ethically appropriate medical practice, clear ethical concerns are raised by parental refusals to have their newborns tested. Similarly, while adults have a right to exercise autonomous decision making, an adult woman's refusal to follow the PKU diet during pregnancy, as is the standard of care, raises ethical issues about competing rights, that is, the woman's right to self-determination or her duty to do what is in her and her fetus's best medical interest. If you permit women the right to choose whether to follow the PKU diet during pregnancy, should those who refuse the diet be held legally responsible for not preventing an avoidable severely disabling disease in their birth-affected children and for creating considerable burden to the state? If you believe that such women should be held legally liable, then shouldn't we also hold liable anyone who knowingly causes any burden to the state, such as individuals who refuse to strap their young children in car seats and crash their cars leaving the children permanently disabled? And, if you agree that the state has a right to limit individual freedoms, how much curtailment is appropriate? Should autonomy be limited only in situations that could result in substantial burden to the state versus those in which the burden falls only on the individual? Should parents of young children, for example, be barred from all high-risk activities, like skydiving?

Other controversial ethical issues exist as well. If we screen for PKU, ought we to screen for other genetic diseases in certain populations? Should normal individuals be screened to determine if they carry detrimental genes that could result in the birth of diseased offspring, such as **sickle cell disease** or Tay-Sachs[11] testing? If such policies are ethically acceptable, are they acceptable only because of genetic involvement, that is, the so-called genetics exceptionalism argument? (The exceptionalism argument is essentially that genetics is so uniquely distinctive that it is functionally exceptional, so policies and practices should account for its distinction from other kinds of biological and medical information and technologies.) Or do the principles supporting the policies extend more broadly to nongenetic situations? For example, should pregnant women be screened for drug and alcohol toxicity known to produce **teratogenic** effects on the fetus?

While some genetic disorders may be clinically diagnosed on the basis of biochemical tests, such as those for enzymatic or chemical levels that can indicate the presence of disease, direct confirmation of genetic variants is done by DNA base testing or biochemical testing of proteins produced by genes. A gene test is essentially an examination of the DNA sequence in a sample of cells taken from blood, bodily fluid, or tissues that reveals whether the genes in question are amplified (too many copies), overexpressed (too active), underexpressed (deactivated or inactive), or absent. Various techniques are used in gene testing, such as using probes or short strings of DNA that have sequences complementary to mutations in question that seek out the complementary sequence, or sequence comparisons between an individual's DNA bases and those in a normal version of the gene. Another type of genetic test, the **karyotype**, can determine whether or not the chromosomes are correct in number and form (e.g., translocation of a part of one chromosome to another with potential loss or addition of genetic material).

Gene tests have various usages, including (1) predictive testing, which is used to identify variants that increase an individual's susceptibility to disease; (2) carrier testing, which identifies the presence of one abnormal allele of a gene, so that reproducing with another carrier gives the couple a 25 percent chance that each offspring will have disease resulting from inheriting the two parental copies of the diseased allele; (3) prenatal testing, which is testing of the unborn offspring to identify chromosomal abnormalities or specific disease; and (4) newborn screening, currently the most widely used type of testing, which determines whether the newborn suffers from a treatable genetic disorder. The goal of newborn screening is to detect a disorder before the onset of symptoms, and potentially irreversible damage, for the purpose of initiating therapy. Newborn screening is mandated by all 50 states, although each state determines which composite of tests are performed, resulting in substantial variability from state to state raising a core question about distributive justice. A few states require as few as three screens (West Virginia, Montana, and South Dakota) while others require up to 30 (North Carolina and Oregon). Ultimately, each state decides which tests are cost-effective enough to be required, though the implicit ethical argument is that the state acts on behalf of infants whose health would be severely compromised by the failure to detect and treat these disorders.

Implementing a screening program to detect carriers (heterozygotes) is arguably a policy matter of a different order of magnitude. The U.S. Public Health Service declared "war" on sickle cell anemia, dubbed the "neglected disease," at a particularly interesting time in U.S. history, specifically on the heels of the Civil Rights movement, an era characterized by profound racial division as evidenced by race rioting in several major cities. Sickle cell anemia is an autosomal recessive disorder affecting roughly 1 of every 500 African Americans, who have severe anemia and pain crises caused by sickle-shaped red blood cells. Heterozygotes number about 1 in every 10 African Americans; they have normal phenotypes and do not develop disease. The high frequency of the disease in the African-American population and the availability of a screening tool spurred a wave of state legislation designed to protect against (prevent) the disease. In 1972, the National Sickle Cell Anemia Control Act went into effect with the goal of addressing inequities in state laws by providing federal funds to both states and private groups involved in sickle cell screening. Interestingly, this first widespread screening for a genetic disease didn't rely on DNA-based techniques, but on more basic technology: observing the shape of red blood cells under the microscope and hemoglobin electrophoresis, which could differentiate the sickled hemoglobin.

The value of *newborn* sickle cell screening is that detection can reduce mortality by 15 percent (affected infants may die quickly from developing pneumococcus infection if undiagnosed). Though well supported by leaders within the African-American community and arguably well intended, laws mandating widespread screening nonetheless are considered by many to have caused more harm than good.[12] Widespread heterozygote screening was fraught with several ethically dubious presumptions. The fact that heterozygotes (individuals with the sickle cell trait, that is, one copy of mutant sickle hemoglobin) would not develop the disease was not well understood at the time. Education programs did not clarify the distinction between trait and disease, and thus avoidable harm resulted. Substantial confusion about the clinical difference between disease and sickle cell trait (i.e., carrier status) led to considerable discrimination against individuals with trait (who would never develop disease, but were believed to inevitably become ill) by the military, employers, and life insurers. Individuals with the trait were forbidden to be airline pilots on grounds that oxygen deprivation at high altitudes would induce disease, which is false. Others were fired from jobs or refused insurance for reasons similarly based on misinformation. In sum, overall, the benefits of testing were unclear and the bases for undertaking the testing ethically unsound.

Some states mandated premarital screening on grounds that couples ought to know they are at risk of birthing a child with sickle cell disease and make reproductive decisions with this knowledge in mind. This led to marriage licenses not being issued to individuals who refused to undergo screening. Mandated premarital screening amounted to the elimination of voluntary autonomous decision making about gene testing, functionally legitimizing coercion. Some states required schoolchildren, who clearly were not of an age to consent, to be tested. Homozygotes typically do not benefit from gene testing, because they've known their diagnosis for years and have a long history of treating it symptomatically. Gene therapy did not then. Today, the technical feasibility of gene therapy is being researched, and therefore it is not yet available to medical practice. Heterozygotes can possibly benefit in making reproductive decisions, but they will not develop disease. (Prenatal detection via amniocentesis or chorionic villus sampling is now available, but it was not then.) Importantly, assurances of medical privacy did not exist for screening. Genetic counseling was hardly ever available. In hindsight, the benefits of testing were not clear, programs to provide ample education and counseling were not in place, and the resulting discrimination as well as mistrust of the medical community on the part of the African-American community may linger still today. Science misapplied to social ails can, in other words, effect long-lasting adverse consequences. Furthermore, the incidence of sickle cell disease, the stated goal of screening, remains unchanged.

Jews during this era were also targets of hatred and ostracism and were targeted for Tay-Sachs screening.[13] Tay-Sachs, a lethal autosomal recessive disorder, is rare but has moderate frequency among Ashkenazi Jews (about 1 in 4,000 births; carriers number about 1 in every 150). Voluntary testing programs were instituted in a few Jewish communities around the country and soon rabbis began counseling couples to be tested premaritally. Testing again would inform reproductive decision making and potentially could spare parents the experience of birthing a diseased child and suffering along with the child until he or she died, usually by the ages of 7 to 9. These programs continued without significant controversy or adversity and are thought to be responsible for the more than 70 percent decline in Tay-Sachs disease. Some believed that when the incidence dropped, rabbis stopped counseling and the incidence increased. Notably, unlike compulsory sickle cell screening, Tay-Sachs screening was voluntary and did not involve misinformation about carrier status or overt discrimination. Noteworthy are the recent findings that a late-onset form of Tay-Sachs exists and the infantile onset form is also prevalent among French Canadians, making Tay-Sachs no longer a "Jewish" disease. Defining disease on the basis of prevalence within certain ethnic, racial, or religious groups may be provisionally useful for increasing indices of clinical suspicion, but doing so threatens to create stigma long after an initial suspicion is proved false, in addition to related problems of definition and self-selection of such groups.

Genetic test results can provide key information to improve health and well-being. Yet even gene tests have their limitations, and it is crucial to understand what they can and cannot tell you. Aside from particular sensitivities and specificities (basically the rate of false-positives and false-negatives associated with a test), a single test does not test for all known mutations, but rather the most common ones. For example, in the case of cystic fibrosis (CF), an autosomal recessive disorder that affects the respiratory and digestive systems and frequently results in death by age 30, more than 1,000 mutations have been identified as associated with the disease, yet the current gene test (as of this writing) examines only 25 variants, those that are known to contribute most to disease. The incidence of CF in the United States is roughly 1 in 30,000, and the mutation is found on the CFTR gene. In 1997, a National Institutes of Health (NIH) panel recommended testing for anyone with a family history of CF, persons with a partner with CF, couples planning a pregnancy, or anyone interested in prenatal testing. As our knowledge of how gene variants and their functionality grow, gene tests are likely to become even more informative, in which case they improve the diagnoses and management of many diseases. For example, knowing the genetic variants associated with over- or undermetabolizing drugs will permit physicians to prescribe medicines they know will give therapeutic benefits and not cause adverse side effects. Drug development will be facilitated by this same information to produce drugs that are maximally effective and safe for specific groups in the population. Furthermore, gene therapy, or the use of genes to treat (if not prevent) both inherited and acquired disease by replacing the dysfunctional genes, may prove to be safe and effective and could become the gold standard therapy for some conditions.

Genetics Testing Today

Today, we know of several hundred single-gene disorders and that these disorders occur rarely affecting only a small fraction of the general population.[14] We also now know that polygenic and multifactorial diseases occur far more frequently and affect nearly all of us, though our understanding of genetic effects is relatively primitive compared with what we expect to know in the future. And, we now know that genes affect virtually all human characteristics, not just diseases.

Experts continue to envision substantial potential for health care innovations based on advances in genomics and related subfields. With genetic tests for roughly 1,000 diseases available (as of this writing),[15] genetic testing now offers one means to betterment. Gene testing can inform potential parents about the risk of having a child with a genetic disease, permitting them to plan for such, and in many cases permitting them to decide to terminate an affected fetus. Genetic testing of newborns permits immediate treatment to reverse the progressive lethal effects of disease, such as in the case of hypothyroidism. Other gene tests promise to promote health by indicating the need for increased medical surveillance (e.g., the use of penicillin prophylaxis in newborns and children once they've been diagnosed with sickle cell disease), a change in lifestyle, or medications that will be safe and effective for an individual patient. Moreover, genetic testing can evaluate the toxicity of chemical compounds and facilitate the development of genetically modified organisms to improve health by removing allergens and adding key nutrients. Use of genetic testing, as decades ago, still presents numerous complex ethical issues involving individuals, families, and society at large.

Given that the major diseases are known to have genetic components, knowledge of an individual's genotype will be increasingly important to elucidate disease development, disease expression, and health maintenance. The mere availability of genetic testing does not ensure that more good than harm will result from its use. Ethically appropriate use of gene testing will require not only compliance with practice standards, but also the physician's ability to accurately interpret a test result and manage patient care accordingly, particularly as the majority of tests will predict relative risk and not the presence or absence of a specific disease. Although the fund of genomic knowledge is growing rapidly, a clinician's ability to predict health risk based on a gene test result in many instances will be variable and limited. A gene test result likely will not indicate *when* a person is likely to become ill during their lifetime, nor the severity of the illness when it strikes, because nongenetic factors significantly influence health status. Determining the medical meaning, or the prognostic significance, of a test result involves assessing variables other than only the presence or absence of a gene variant, such as an individual's present, past, and anticipated exposures, other biological processes involved, and even one's socioeconomic situation, such as the extent of health benefits and whether one can pay for treatments not covered by insurance, for example, treatment for alpha(1)-antitrypsin disease. Enzymatic replacement treatment for alpha(1)-antitrypsin disease, which over time reduces symptomatology, improves health measures, and prevents adverse effects of disease, costs roughly $31,000 annually. Symptomatic, that is, nongenetic, treatment costs roughly $20,000 annually. Insurers prefer to pay for acute events resulting from disease-related symptomatology, because on an event-by-event basis it is cheaper than chronic care, even though in the long run chronic care is both cheaper and more effective given that it's preventive.[16]

Despite the overwhelming homogeneity in the human genotype, the 0.1 percent difference may prove so divisive as to create a genetic underclass of individuals whose genetic lot imposes societal cost burdens. In efforts to ensure that all humanity can benefit from genomic advances, the United Nations Education, Scientific and Cultural Organization, in 1997, promulgated the *Universal Declaration on the Human Genome and Human Rights.*[17] This document commands international commitment to advancing genomics only through the preservation of the fundamental sanctity of human dignity, freedom, and fundamental human rights. Of primary concern to international bioethics committees is that genomic testing improves the world's health rather than increase existing health disparities.[18]

The ethical concerns about the use of genetic testing are vast and complex. Several questions should be pondered: Ought tests be used if no proven disease treatment or strategy to mitigate risk exists, even if individuals desperately want to know? Should genetic tests be regulated differently than other medical tests? For example, ought consumer nutrigenomic tests be regulated differently than mammography? Should regulations exist for all types of genetic defects or just some (e.g., are tests for poor eyesight the same as tests indicating a tendency to baldness)? Should tests be used on in vitro embryos to ensure that those that test positive are not implanted?

What if one wants to know their genetic profile, not for themselves, but for future offspring? Can individuals be protected from harm? What if years later the genetic test results offer additional (say, negative information, and in fact something so negative the individuals never would have consented to it)? For example, Apo E4 identification can be useful in determining and lessening cardiovascular risk through diet and exercise. However, APO E4 can also indicate increased risk of Alzheimer disease. With the ever-burgeoning medical knowledge, it is impossible to plan for such consequences in the future. Primary ethical concerns arising in the provision of genetic tests are as follows:

- Who owns genetic information?
- What constitutes fair use of genetic information?
- How can we ensure that genetic tests will be evaluated for accuracy, reliability, and utility before they are commercially available?
- Can informed consent for research and clinical testing be assured, given the current low level of genetics literacy and difficulties in enforcing nondirective counseling standards when interpretation is required?
- Can we ensure that all testing is voluntary?
- Should parents always have the right to have their underage children tested?
- Can we ensure the privacy of genetic information?
- Can we ensure the appropriate handling of biological samples?
- Can we ensure the appropriate introduction and use of genetic tests and genetic data?

The bulk of ethics research has identified and clarified the potential for harm, which in turn has led to recommendations, laws, and policy intended to prevent or mitigate that potential (e.g., state anti–genetic discrimination legislation or NIH consent guidelines). A smattering of high-profile cases illustrates some of the types of ethical quandaries that challenge both definitions and limits.

Duties and Obligations

What should the states' obligations be with regards to genetic testing, particularly in light of competing needs for limited resources? In all cases, beneficence competes with other state demands. Furthermore, how can the state ensure maximal benefit while realizing that mandatory screening adopted based on public health rationale does not necessarily mean that everyone equally benefits from testing? For instance, the type of genetic test reporting required by the state could lead to devastating consequences, such as events in a recent case in California. California's newborn screening reporting requirements do not require reporting actual values, but rather only interpreted values (abnormal results). As a result, one newborn's screen was interpreted as "normal," but actual values showing for congenital hypothyroidism were not reported to the pediatrician until several weeks later, at which point irreversible harm resulted. Had the physician received the actual values early, preventative treatments could have been prescribed. Parents sued the state hoping to recoup damages that would help pay for expensive lifelong care, but the ruling found that the state not liable.[19] A similar recent case occurred in which a new mass spectrometry screen was being piloted. One newborn was offered the screen, parents consented and the infant was identified with a metabolic disorder that was successfully treated; the child was now normal. An infant in a different state, where this screen was not mandated, went undiagnosed for nearly two years, with the end result being serious irreversible disability for the rest of his life.[20] Ethical issues arising from inconsistent newborn screens among states are a current hot topic. A related concern is the fact that health care consumers, particularly those in areas using multiplayer systems of care, have widely differing levels of access to tests and interventions. In the United States, changes to one's health insurance can create substantial potential for harm if insurers refuse to pay for expenditures associated with a "preexisting" (genetic) condition or impose premium increases that make treatments (even copays) unaffordable. In other parts of the world where individual freedoms are restricted, by law or custom, the use of genetic testing may be even more ethically problematic.

Genomic medicine challenges physician duties in the context of doctor-patient relationships. Physicians have a duty to protect a patient's confidentiality. This duty is reinforced by the Health Insurance Portability and Accountability Act of 1996 (HIPAA),[21] which established standards for the privacy and security of health information. However, such a duty is not absolute. Doctors are obligated to override patient confidentiality to report infectious disease (i.e., tuberculosis) and threats of violence (as in the Tarasoff decision).[22] Whether genetic risks warrant similar exceptions, and if so, when, remains an important clinical concern. Several advisory committees have recommended that physicians be permitted to breach confidentiality to warn third parties only as a last resort to avoid serious harm. This question reached the Florida Supreme Court in 1995 when a woman sued her mother's physician for failure to warn her about the genetic transmissibility of her mother's medullary thyroid carcinoma, arguing that had she known the diagnosis was hereditable she would have taken preventive measures and herself avoided the disease. The court ruled that the physician did have the duty to warn the patient (the mother) based on the standard of care, though the duty did not extend to the daughter requiring a breach in patient confidentiality.[23] What defines "serious" harm, however, is clearly different for different people, and depends on environment, heredity, psychological, and other factors of their situation.

Furthermore, although HIPAA's primary concern is electronically stored and transmitted health information, its privacy protections extend also to data maintained or exchanged in other mediums, namely paper records, faxes, and even oral communications. Even though HIPAA contains privacy provisions to safeguard genetic information, it focuses so entirely on the individual that it fails to take into account that an individual's genetic privacy needs extend to blood relatives and inevitably are tied to specific contexts that can create conflicts between "what" is known and "when" it becomes known.

There are other questions of physician obligations including, but not limited to, whether a physician is ever obligated to offer a genetic test. In the absence of appropriate indications and established standards of care, many argue that physicians have no special duty to offer a specific test. Conversely, its widely agreed that physicians do have an ethical obligation to be knowledgeable about genetic disorders that occur with increased frequency in certain populations, such as sickle cell anemia in African Americans, Thalassemias in Mediterranean and Southeast Asian groups, and Tay-Sachs in the Ashkenazi Jewish and French Canadian population. This knowledge is used to increase indices of suspicion within their patient population. Physicians have an ethical obligation to provide nondirective counseling in support of a patient's voluntary decision making.[24]

Voluntary Autonomous Decision Making

Essentially, there is universal agreement at the national and international levels that the choice about whether to undergo a genetic test must be voluntary and without coercion. Exceptions involve individuals who are deemed incompetent and children. Practice standards and policy recommend that children should be tested only if there is proven medical benefit in doing so. In practice, the benefits of childhood testing depend on the family's ability to access services indicated by a test result. Furthermore, though the law supports parental prerogative, parents may not necessarily act in their child's best interest. Ethical conflicts about what is truly in their best interest could easily arise when parents authorize testing and their child does not want (or in the future decides they did not want) to have his or her genetic status known or available. Moreover, existing law may be insufficient to protect the interests of children against abuses instigated by parents or the government.

The potential to coerce individuals into testing should not be minimized. Third parties with a financial stake in a genetic test result have tested individuals without their consent, and future instances are quite possible. Recently, the Burlington Northern Santa Fe Railroad tested 35 employees seeking disability compensation for job-related carpal tunnel syndrome for mutations associated with hereditary neuropathy. The employees learned they were tested by accident. The U.S. Equal Employment Opportunity Commission halted the practice, and a class action suit was settled out of court. In point of fact, the scientific basis for the testing undertaken was inappropriate.[25] Hereditary neuropathies occur rarely in the population and while carpal tunnel syndrome can be a symptom of the condition, it is one of many symptoms that characteristically arise well before midlife, which is when the workers developed the carpal tunnel syndrome.

The government also has a stake in the health of its citizenry. It is plausible, therefore, that it could require genetic testing to limit its financial burden. In 1992, the state of Colorado considered mandating prenatal Fragile X screening of prospective parents with the belief that large-scale screening, coupled with reproductive counseling, would result in fewer births of affected individuals and thus reduce the state's financial burden arising from the care of such individuals. Studies (though flawed in design and result) indicated that each person with Fragile X costs the state $1,609,852.63 more than what is expended for "normal" children. Extrapolating that number to a national burden of $280 billion, researchers argued that mandated screening could relieve financial burden by 50 percent.[26]

Other types of state action are equally imaginable. To avoid state expenditures on unsafe and ineffective medications, a government might require those on state assistance to undergo pharmacogenetic testing (i.e., a test to determine whether a medicine will be safe and effective for an individual person) as a condition of having their medication paid for. Insurers might require such testing for similar reasons. Here, a concomitant ethical concern is that such individuals might unwillingly receive personal genetic information they never consented to have tested and never wanted to know. While the risk that information obtained from a genetic test comes to mean more than what one initially consented to know exists for anyone undergoing testing, it is particularly problematic in cases in which testing is directly or indirectly coerced.

Genetic Discrimination

Genetic discrimination, defined as the exclusion of entitlements and social benefits based on an individual's genotype (known or presumed), was briefly discussed earlier in this chapter in regards to sickle cell screening programs in the 1970s. It remains a potent threat in light of the increasing availability of genetic tests and the danger that personal genetic information may end up in the possession of unauthorized third parties. A genetic test result can have profoundly adverse effects on an individual and their family members. A genetic diagnosis might affect a person's ability to perform a job that could not be accommodated for in reasonable ways. For example, an individual with recurrent and untreatable cardiac arrhythmia that leads to loss of consciousness, owing to an inherited ion-channel defect, would be ineligible to work as a long-distance truck driver. An individual at risk for chronic beryllium disease, given sufficient exposure, may be "protected" out of jobs that increase his or her exposure and risk of developing disease. In both cases, individuals could be barred from their preferred occupation, raising the question of how to balance competing rights: an individual's right to self-determination (to choose for themselves which risks he/she is willing to incur) and an institution's right to maximize their own best interest or right to protect individuals against harm to themselves or the public (i.e., the public from dangerous driving or an individual from increased of serious disease). Balancing competing interests is complicated not only by the potential for greater good (say, a brilliant chemist with inherited susceptibility for chronic beryllium disease, who if permitted to work in his risky, but preferred, occupation might make a discovery of enormous benefit to society), but also by the uncertainty about the likelihood of various outcomes associated with a test result. For example, should an asymptomatic individual with a predisposing, but incompletely penetrant, mutation for the ion-channeling defect be as restricted as a symptomatic individual with complete penetrant mutation? Framing the risk-benefit question is far from simple.

Determining whether it is ever ethically acceptable to limit an individual's right to societal entitlements or benefits is likely to remain contentious. Furthermore, discrimination is typically covert, not overt, and therefore difficult to identify and remedy. For example, individuals often disclose private information in the context of interpersonal trust, such as friend to friend. Conflict of interest in such disclosures exists when one person has competing obligations, for example, obligations to the friend but also obligations to their insurance agency employer. If the recipient (the insurance agent) determines that their primary obligation is to their employer, then information obtained in confidence is likely to be used against the friend. Here a disclosure is not entirely consensual, because the donor may be wholly unaware that their confidential information could be shared without their prior authorization. Other types of involuntary disclosures are also possible. In small rural areas, one's family history may well be public knowledge. Equally, if not more troubling, may be the potential for self-imposed exclusions based on personal beliefs about self-worth and entitlement connected to one's understanding of their genetic test result. Individuals may deny themselves education, marriage, children, and so on, because they believe their test result indicates that they are unworthy. Given that we all harbor 8 to 25 mutations that if present in two copies would cause disease, we are all theoretically vulnerable to genetic discrimination at any point in our lives.[27]

Informed Consent

The ethical constraints that apply to any medical test apply also to genetic testing, though ensuring informed consent is complicated by the inherent complexity of genetics, as well as by the fact that the patient is not only the individual, but also his or her blood relatives. Ensuring the genetic literacy of all is a major unmet need. Whether individuals are told all that they need to know may be compromised by the practitioner's information and level of understanding (also an important as yet unmet need). Whether individuals truly understand the information provided also may be questionable, as the consent process itself is not foolproof. Understanding genetic risk is particularly challenging, because the tendency to simplify to ensure comprehension can backfire by reinforcing incorrect notions of biological determinism. From an ethical standpoint, conveying information via interpreters is especially problematic for several reasons. Interpreters are not required to be trained in genetics, or genetic counseling standards, nor in the ethics or cultural norms associated with nonnative speakers. Lacking such training, interpreters may be more likely to unintentionally use language that isn't neutral in meanings. Often children are used to interpret for parents, raising a host of issues, including patient privacy, the appropriateness of involving children in the intimate issues raised, and the effect of such exposure on a parent's future relationship with the child.

Balancing the privacy rights within families may be contentious. Inasmuch as genetic testing can require family history information, there is potential for harm to the patient (the would-be testee) if family members are unwilling to participate. In 1998, a participant in a clinical genetics research study provided her father's family history information, as per the study protocol. But the father blocked the researcher's access to his medical information, arguing that he did not consent to disclosure, and disclosure without his consent amounted to a right to privacy violation. The question of whether disclosure of family history information requires the consent of each family relative remains uncertain, as it does when consent applies to deceased individuals. And if it does require such consent, how would it be achieved? Moreover, in the case of identical twins where one wants to know and the another does not, determining whose right merits priority is difficult, particularly in light of the fact that it's highly improbable that both rights could be exercised and adequately guarded.

Just Distribution of Benefits

Each of us has a stake in how genetic research unfolds and the resulting technologies are applied. These stakes are particularly high because of the magnitude of unknowns involved. For example, high-speed easy chip testing for information A might result in the unanticipated generation and release of information B and C as well. Active involvement by stakeholders may go far toward fostering just outcomes. In 2003, a court established a patient's *right to access,* which represented a landmark in patient advocacy. Canavan disease patients won the right to remain active participants in the advancement of research and access to the public.[28]

The value of a genetic test lies not in the test itself, but rather in what we, as individuals and society, choose to do with that result. Assuring equitable access to the benefits of genetic testing remains a bonafide concern. Ongoing involvement by international organizations will be required to ensure that individuals in developing countries are not ultimately harmed by such testing and that they, too, can reap the benefits of genetic technologies and engage in setting policy. What we do with genetic information arguably will have a far greater impact on health status than the genetic information itself. Whether individuals use test results to alter their behavior to improve their health remains uncertain, especially given evidence that, despite high levels of motivation, psychological and other factors have the potential to drive people to act in ways that are contrary to long-term health goals.

Many people feel that a gene test is inherently valuable (the knowledge of your genetic status is a benefit independent of any actions you would take on the basis of that knowledge), but others believe that the benefits of a test are dependent on what one can do with the information (or test result). For example, testing positive for a *BRCA* (breast cancer) gene, which increases the likelihood for developing breast cancer, does not mean that one necessarily will develop the disease, nor does testing negative for a particular *BRCA1* mutation mean that breast cancer will not develop from other causes, including but not limited to yet-to-be-discovered genetic changes or gene-environmental factors.

As genetic technologies evolve, the standard of patient care undoubtedly will change to accommodate the benefits of genetic information. This process will continue to generate innumerable ethical concerns (just some of which are noted in this chapter). How we as a society decide to use genetic tests will affect health care financing, underwriting health insurance, state and federal health care programs, and determinations of who pays, if anyone, for the uninsured. Will genetic tests only be available to those who can pay out of pocket, which in essence creates a two-tiered health care system, or will the state mandate the use of genetic tests to improve the cost-effectiveness of state-financed health care?

Conclusion: Threads of Teleology

This discussion illustrates the concomitant drive for new genetics knowledge for its own sake and for the application of such knowledge to improve humanity's lot. Indeed, despite the shameful injustices resulting from past applications, most geneticists still believe that genetics ought to serve human needs and remain committed to using new genetic knowledge for the betterment of society. Proof of the commitment to ensuring good over harm arguably can be seen in the government's decision to use a substantial proportion of the NIH's budget to establish its ELSI (Ethical, Legal, and Social Implications) Division[29] to anticipate the potential for adverse outcomes, to research that potential, and to recommend policies and strategies for mitigating or even preventing harm.

When the Human Genome Project began, the genome was thought to contain 100,000 genes. With its completion in 2003, we now know that the total number is closer to 20,000 to 30,000 genes. Having the so-called book of human life in hand, the hope is that we can prevent the miseries of genetic "bad luck" in the future. The genetic lottery is hardly a just system and its distribution is certainly inequitable. Moreover, when gene variants can dramatically ravage a human life, it is hard not to think that there are bad genes and good genes. Our culture is so subsumed with norms about what defines good and bad that many worry that illuminating genetic influences on outcomes that are deemed one or the other will only undermine free will and personal responsibility. Individuals can blame their lot on their genes, thereby trying to absolve themselves of any responsibility for any unseemly act or status. And yet, genes themselves are neither good nor bad. Nor do they, except in extremely rare situations like single-gene dominant disorders, single handedly cause the outcomes (phenotypes) with which they are associated with, because typically numerous other factors influence their effect (phenotype). As emphasized earlier in this chapter, even in the case of autosomal dominant disorders with complete penetrance, where a predispositional test could tell you *whether* you have the gene and thus whether you will develop disease, a test result cannot tell you *when* it will develop, or its severity when it manifests. In other words, genotype does not precisely determine phenotypes.

Efforts to harness genetic knowledge and leverage it to improve various facets of life throughout the history of scientific discovery reveal punctuated surprises, both intended and unintended societal harms and benefits, as knowledge forges ahead. Despite our best efforts to anticipate and resolve ethical quandaries arising from using genetic technologies, unforeseen dilemmas will continue to emerge. Ethics, though, has a pivotal role to play in ensuring that we achieve our goal of ensuring that more good than harm comes from applying genetic innovations, particularly in guarding against the reification of genetics as the arbiter of ethical decisions.

All science is bound up with ethics in some way or another, simply because the value of science is in large part what we can do with it (namely, how it can be applied to what end) and whether we ought to use it as such simply because we can. While this fact holds true for any science, it is arguably most particularly true for genetics. Genetics is bound up with who we are and how our lives unfold. That is not to say that genetics determines who we are or determines our life narratives. The history of science, replete with theories, evidentiary proof, and epistemological challenges, is told as the sequential discovery, validation, or refutation, and subsequent justification of truth. Rarely, if ever, has the utility and moral implications of scientific advance been so questioned concomitant to scientific pursuit, such as is the case with genomic science.

Notes

1. His findings were lost but then rediscovered and laid out in Morgan, T. H., Sturtevant, A. H., Muller, H. J., and Bridges, C. B., *The Mechanism of Mendelian Heredity* (New York: Holt, 1915).

2. Mange E. J., and Mange, A. P., *Basic Human Genetics* (Sunderland, MA: Sinauer Associates, Inc., 1994).

3. Morgan, T. H., Sturtevant, A. H., Muller, H. J., and Bridges, C. B., *The Mechanism of Mendelian Heredity* (New York: Holt, 1915).

4. Mange, E. J., and Mange, A. P., *Basic Human Genetics* (Sunderland, MA: Sinauer Associates, Inc., 1994).

5. Galton, F., "Eugenics: Its Definition, Scope, and Aims," in *Sociological Papers* (London: Macmillan, 1901), 45–50.

6. Galton, F., "The Possible Improvement of the Human Breed Under the Existing Conditions of Law and Sentiment," *Nature* 64 (1901): 659–65. Galton, F., *Essays in Eugenics* (London: Eugenics Education Society, 1909).

7. Davenport, C. B., *Heredity in Relation to Eugenics* (New York: Holt, 1911).

8. Galton, F., "The Possible Improvement of the Human Breed Under the Existing Conditions of Law and Sentiment," *Nature* 64 (1901): 659–65. Galton, F., *Essays in Eugenics* (London: Eugenics Education Society, 1909).

9. Miescher, F., "On the Chemical Composition of Pus Cell," *Hoppe-Seyler's medizinische-chemische Untersuchungen* 4 (1871): 441–60. Abridged translation in Gabriel, M., and Fogel, S., eds., *Great Experiments in Biology* (Englewood Cliffs, NJ: Prentice Hall, 1955). Original work in *Die histochemischen and physiologischen* (Leipzig: Vogel, 1897).

10. Paul, Diane B., *The Politics of Heredity Essays on Eugenics, Biomedicine and the Nature-Nurture Debate* (Albany: State University of New York Press, 1998).

11. Sickle cell hemoglobin results from a change in 1 of 146 amino acids in the β globin; a substitution of valine for glutamic acid at the sixth position of the polypeptide. Homozygosity for this mutation causes sickle cell disease, a severe hemolytic condition characterized by the tendency of red blood cells to become grossly abnormal in shape (sickle-shaped) under conditions of low oxygen tension. The disease causes anemia, crises from hemolysis or vascular occlusion, chronic leg ulcers and bone deformities, and infarcts of the bone or spleen. Sickle cell disease occurs most frequently in equatorial Africa, and less so in the Mediterranean area, India, and countries to which people from these regions have migrated. The disease occurs in roughly 1 out of every 600 African Americans. Tay-Sachs is an autosomal recessive neurological degenerative disorder caused by a deficiency of hexosaminidase A and affects roughly 1 in every 360,000. The carrier frequency is far more common in Ashkenazi Jews and also prevalent among French Canadians. Disease typically develops by the time a child is 7 to 9 years of age. Affected children develop blindness, regress

both mentally and physically, and usually die in childhood, although in the 1990s a form with onset in late adolescence and early adulthood was discovered wherein affected individuals lived into adulthood.

12. Faden, R., Geller, G., Powers, M., eds., *AIDS, Women, and the Next Generation: Towards a Morally Acceptable Public Policy for HIV Testing of Pregnant Women and Newborns* (New York: Oxford University Press, 1991).

13. McQueen, David, "The Diffusion of the Innovation of Genetic Screening: Tay Sachs Screening in Baltimore and Washington, D.C.," dissertation, 1973.

14. http://www.ornl.gov/sci/techresources/Human_Genome/medicine/assist. shtml; Online Mendelian Inheritance in Man, http://www.ncbi.nlm.nih.gov/entrez/query.fcgi?db=OMIM.

15. www.genetests.org.

16. http://www.musc.edu/alphaoneregistry/investigators/recent_article_synopsis. html; Alkins, S. A., and O'Malley, P., "Should Health-Care Systems Pay for Replacement Therapy in Patients with Alpha(1)-Antitrypsin Deficiency? A Critical Review and Cost-Effectiveness Analysis," *Chest* 117, no. 3 (March 2000): 875–80; http://www.aetna.com/cpb/data/CPBA0145.html.

17. United Nations Educational, Scientific, and Cultural Organization, The Universal Declaration on the Human Genome and Human Rights, http://portal.unesco.org/shs/en/ev.php@URL_ID=2228&URL_DO=DO_TOPIC&URL_SECTION=201.html (accessed March 2005); The World Health Organization, Genomics and World Health, http://www3.who.int/whosis/genomics/genomics_report.cfm (accessed March 2005).

18. Daar A., and Mattei J-F., Draft World Health Organization (WHO) guidelines on bioethics. *Nature*; World Conference on Science, http://www.nature.com/wcs/b23a.html (accessed March 2005); http://www.who.int/genomics/publications/en/; and Report of WHO on Proposed International Guidelines on Ethical Issues in Medical Genetics and Genetic Services. Genetics Education Center, University of Kansas Medical Center, Ethical, Legal, Social, Implications of the Human Genome Project, links to policy papers on specific issues, professional societies, http://www.kumc.edu/gec/prof/geneelsi.html (accessed March 2005); U.S. Department of Energy Office of Science, Human Genome Project Information, "Evaluating Gene Tests: Some Considerations," http://genome.gsc.riken.go.jp/hgmis/resource/testeval.html (accessed March 2005); Human Gene Organization Ethics Committee, "HUGO Statement on Benefit Sharing, April 9, 2000," *Clinical Genetics* 58, no. 5 (2000): 364–66, http://www.gene.ucl.ac.uk/hugo/benefit.html (accessed March 2005); HUGO Ethics Committee, "HUGO Urges Genetic Benefit-Sharing," *Community Genetics* 3, no. 2 (2000): 88–92; Knoppers B., and Chardwick R., "The Human Genome Project: Under an International Ethical Microscope," *Science: New Series* 265, no. 5181 (1994): 2034–35; Clayton, E. W., "Ethical, Legal and Social Implications of Genomic Medicine," *New Journal of Medicine* 349, no. 6 (2003): 562–69.

19. *Sierra Creason v. Department of Health Services*, no. SO63167, Supreme Court of California, 18 Cal. 4th 623; 957 P.2d 1323; 76 Cal. Rptr. 2d 489; 1998 Cal. Lexis 4038; 98 Cal. Daily Op. Service 5477; 98 Daily Journal DAR 7619.

20. "Testing Fate: A Drop of Blood Saves One Baby; Another Falls Ill. Inconsistencies in Screening Mean Rare Diseases Go Undetected and Untreated," *Wall Street Journal*, July 17, 2004, A1 column 1, A13 columns 1–21.

21. HIPAA went into effect in 2004.

22. *Tarasoff v. Regents of University of California*, S.F. No. 23042, Supreme Court of California, 17 Cal. 3d 425; 551 P.2d 334; 131 Cal Rptr.14; 1976 Cal. Lexis 297; 83 A.L. R. 3d 1166, July 1, 1976.

23. http://www.genelaw.info/pages/casedetail.asp?record=88 and http://biotech.law.lsu.edu/cases/genetics/Threlkel.htm.

24. Genetic Counseling Resources, Genetics Education Center, University of Kansas Medical Center, http://www.kumc.eud/gec/prof/gc.html; Juengst, E., "Caught in the Middle Again," *Genetic Testing* 1, no. 3 (1998): 189–202.

25. http://www.mindfully.org/GE/GE4/Railroad-Workers-Genetic-Defects 8may02.htm; http://www.sph.unc.edu/nccgph/phgenetics/burlington.htm.

26. Billings, P., and Hubbard, R., "Fragile X Testing: Who Benefits? School-Sponsored Genetic Screening Program Raises Specter of Discrimination," *GeneWatch* 9, nos. 3–4 (January 1994): 1–3; Tassone, F., Hagerman, R., et al., "Transcript of the FMR1 in Individuals with Fragile X Syndrome," *American Journal of Medical Genetics* 97, no. 3 (2000): 195–203; International Conference Proceedings, Snowmass/Aspen, CO, 1992, National Fragile X Foundation, Denver, CO, 1992; Murray, J., et al., "Screening for Fragile X," *Health Technology Assessment* 1, no. 4 (1997); Gene Watch, Council for Responsible Genetics, "Fragile X Screening Proposed in Colorado," February 1993[0]; [0]http://www.ncchta.org/execsumm/SUMM104.htm.

27. Marchant, J., and Day, M., "DNA—Celebrating 50 Years of the Double Helix," *New Scientist Magazine*, May 20, 2000.

28. *Medical Research, Law and Policy Report* 3, no. 11 (June 2, 2004): 440–50, copyright Bureau of National Affairs, http://www/bna.com; http://raredisease.about.com/library/weekly/aa011201a.htm.

29. National Institutes of Health, National Human Genome Research Institute, "The Ethical, Legal and Societal Implication (ELSI) Research Program," http://www.genome.gov/; http://www.genome.gov/PolicyEthics.

CHAPTER 2

What Is Ethical Analysis?

Moral virtue is acquired by repetition of the corresponding acts.
—Aristotle, *Nicomachean Ethics* (Book II, Chapter 1)

Pleasure in doing virtuous acts is a sign that the virtuous disposition has been acquired.
—Aristotle, *Nicomachean Ethics* (Book II, Chapter 3)

A central goal of this book is to improve your ability to identify and analyze ethical issues. Without some basic understanding of what characterizes ethical issues, as opposed to factual disputes, and what types of reasoning are used to evaluate the consistency, coherence, and overall strengths of claims embedded in arguments, such a goal is not likely attainable. To that end, this chapter provides the basic knowledge and skill needed to aptly delve into the subsequent topics covered in the remaining chapters. With a focus on how to identify genuine ethical dilemmas and what it means to clarify and resolve them, the following concepts of moral theory, applied ethics, ethical concepts, and principles are introduced.

Moral theory has long been the province of erudite ethical theorists and religious scholars, and often it has been considered too abstract to be of much relevance to daily life. Its rich history, though, offers rules and reasons that define good and bad actions and explains what it means to behave responsibly toward one another. Ethical practices represent the application of ethical principles to questions of conduct. Engaging in ethical analysis involves (1) clarifying reasoning (implicit and explicit) used to justify a position, (2) identifying logical inconsistencies, and (3) determining the strength of arguments and their underlying rationale, for the purpose of determining the right or fair thing to do, understanding that the right—that is, the morally superior—thing to do is not necessarily the equally fair thing to do.

How Do You Know an Ethical Issue When You See One?

In reality, most of us get by fairly well in life without having to resort to philosophical analysis for help. We know that it's wrong to lie and to steal. We also know that it's good to help others, particularly those who are less fortunate than ourselves. But if you ask people how they know what is right or wrong, they typically answer, "I just do," or "It's what my gut tells me," or "I feel strongly that it's right." None of these explanations would pass muster in a philosophical analysis, but they seem adequate to guide us in navigating moral confusion in our day-to-day activity.

We run into trouble, however, when we're not sure about what the right thing to do is. For example, you might experience an uncomfortable feeling in the pit in your stomach if you're doing (or want to do) something you know you should not do. The queasy feeling suggests ambivalence about what the right decision is. Both courses of action seem correct, yet together they contradict one another. What if there's little risk in getting caught if you choose the wrong course of action? Does that tip the balance between your choices?

We also run into trouble when holding a position that is inconsistent with norms or principles that have guided us previously. Imagine that you believe stealing is *always* wrong and should be punished accordingly. When asked what the appropriate punishment is for a 6-year-old child found stealing food from the cafeteria, you might pause to consider whether your usual recommendation for stealing, school suspension, ought to apply to this child. You might consider whether the child who looks malnourished may be desperate for food, because his parents cannot provide adequately for him. If so, you may be inclined to use a different standard of fairness than the one you usually apply. Here you are considering conflicting principles of fairness when deciding which course of action is correct (or ethically defensible) and why. In trying to determine what is fair, you might ask yourself for the justification to make an exception for the young or starving child. Should this exception be generalized to any child similarly situated? Does it matter *what* the child is stealing? Would you allow the child to go scot-free if he or she was stealing guns with the intention of shooting people? Does age matter? If a teenager were involved, would the teen be more culpable than a much younger child?

When our consciousness acknowledges a moral conflict, resolving that conflict requires identifying and evaluating reasons, justifications, and implications. An ethical dilemma, in other words, is a situation in which an agent T believes he/she morally ought to, but cannot, do both action A and action B. Agent T cannot do both actions because B is opposed to A in some way or because situational features prevent doing both actions.

Classic examples of such dilemmas are as follows. A person makes two promises with the intention of keeping each, but then discovers that it is impossible. For example, a physician consents (i.e., promises) to maintain the confidentiality of test results in the context of research they are conducting. Contradictorily, consider the possibility that one of the research participants is also a patient in the practice and the physician obtains research results that may be beneficial to the patient. While the research conditions forbid the physician from using the results to provide clinical care, the physician is arguably in a bind because he has promised to maintain confidentiality of research findings and to do no harm to the patient. If withholding significant clinical information and not acting on it to improve the patient's health status amounts to harming the patient, then a conflict (ethical dilemma) exists. Another example is the famed hypothetical life boat situation in which supplies permit only one of the two passengers to survive. In other words, one life, not two, can be saved, raising the question, what is the right thing to do? Should one sacrifice their life for the sake of the other or should both split supplies with the high likelihood that neither will survive. Is it fair to sacrifice one for the benefit of the others? Is it right to save one life for the greater good of society, in which case greater societal value may be placed on one person's life over the other person's life? For example, one might consider that the life of a young physician who specializes in infectious diseases might be more socially valuable than that of an older homemaker without children. That is, if one person's life ought to be saved, who's should it be and why? In both examples, the general principle we uphold, that innocent lives must always be saved, enjoins incompatible actions. This characteristic is precisely what distinguishes practical dilemmas (should I exercise or not?) from moral dilemmas (who's life is more valuable?).

When we engage in ad hoc justifications for our opinions about what's right or wrong, we generally raise a host of normative issues, explicitly or implicitly, that typically creates more confusion than clarity. Identifying and analyzing claims, and their normative underpinnings, enable us to clarify our thoughts to make explicit rules and standards that are implicit in our judgments. In situations in which rules and standards are found to conflict, we try to resolve the conflict by suggesting modifications that nonetheless remain coherent with our underlying concepts of morality, duty, responsibility, justice, and happiness. When cogent or plausible reasons exist for opposing positions, we want to know not only what is the "right" thing to do, but also which of the competing principles (what ought to be done and why) is most compelling. The latter moves one from the sphere of practical reasoning into the sphere of theoretical reasoning about rules, principles, and generalizability of norms of conduct.

For example, several theories on ethical decision making are likely to disagree about the right thing to do, because they define *the right thing to do* differently. The theory of consequentialism (discussed later in this chapter), for example, places greater emphasis on the well-being of the group than that of any individual group member, whereas the theory of autonomous decision making places greater emphasis on the individual than the group. According to the former, the right thing to do is whatever maximizes the benefit to the group (i.e., creates the most benefit for the greatest number of people). According to the latter, the right thing to do is that which most expresses one's freedom to voluntarily choose for oneself (i.e., one's right to self-determination). As a practical matter, determining the right thing to do involves modifying theories to apply to specific situations to analyze the reasoning for each position.

Clearly, any civil society is governed by rules, laws, mores, and social norms. Moral theories and theories of obligation, for example, explain why obligations are binding and independent of social norms or mores. Generally, we know and feel that we should do what's right and not do what's wrong. Sometimes, what we know we ought to do can conflict with our personal inclinations to do otherwise. For example, I know I ought to pay my debts whether or not I want to do so. Or, I know I ought not to steal or kill, even if I desperately want to do it. Similarly, social norms govern what's acceptable in certain social groups. Such norms indicate that certain acts are right or good. However, one set of norms governing one society, by definition, do not necessarily apply to another society, and thus they do not necessarily define what is right or good in every context. For example, norms of some social groups condone, if not encourage, infanticide, whereas in the United States infanticide is an illegal, criminal, and morally abhorrent act. In some parts of the world, infanticide is practiced when families birth girls but only want sons, or when babies are born after the family has reached its desired size. The slaughter of helpless infants would seem to be universally and absolutely immoral, except possibility under conditions where death to the infant is a lesser harm or greater good than the alternative(s). The point, however, is that the mere fact that infanticide is customary does not make the practice morally justified. Similarly, the sanctioned practice of maiming slaves before the Civic War did not make this cruel and vindictive treatment morally appropriate or even justified.

Fact versus Value: Description versus Prescription

Generalizing from actions in specific situations to rules of good conduct, responsibility, or entitlements (rights) might appear similar to scientific generalization, but there are important distinctions. In science, observed facts or patterns of events within the context of testing a particular hypothesis permit the creation of theories that both explain the causality underlying the observed phenomenon and predict future occurrences. In other words, a scientific theory gives us a provable reason for explaining what did happen, determining what is happening now, and predicting what is likely to happen. The proof is simply in the observed facts that describe the existence and changes of things in the world.

In philosophy and ethics, this distinction is referred to as the fact-value or is-ought distinction. The concept is used to distinguish arguments that can be claimed by reason alone from those whose rationality is limited to collective opinion. In a slightly different characterization, it is the distinction between what "is" (i.e., can be discovered and proved by science, philosophy, or reason) and what "ought" to be (i.e., a judgment that can be defended rationally but not proven true or false). The terms "positive" and "normative" represent another formulation. Positive statements display implicit claims to facts. For example, water is composed of two hydrogen atoms and one oxygen atom. Normative statements, on the other hand, contain claims to values or norms. For example, access to water, a necessity for life, is a fundamental right.

David Hume (1711–1776), the Scottish philosopher credited with establishing the fact-value or is-ought distinction (in his Treatise), argued that the logical independence of fact and value, what is and what ought to be, are distinct such that ethical claims cannot be drawn from natural facts. In other words, he argued that "ought" cannot be derived from "is." Normative statements do not, that is, follow from positive ones. A subsequent English philosopher, G. E. Moore, propounded a related but different problem, which he coined the naturalistic fallacy. In his 1903 book *Principia Ethica*, Moore argues that attempts to prove an ethical claim by appealing to a definition of "good" (or any of its "natural" properties, such as pleasant, desirable, right, and so on) constitute a formal (or logical) fallacy. The naturalistic fallacy refers to the logical mistake that occurs when one claims that what is natural is inherently good or right and reciprocally that what is unnatural is bad or wrong. Notably, many contemporary ethicists have challenged the validity of these assumptions. Nonetheless, the fact-value distinction has been used in modern social science.

Scientific Method and Informal Reasoning

Evaluating scientific claims involves analyzing the accuracy of the data provided and the extent to which it supports derived conclusions. Evaluating the scientific reasoning underlying the claims is equally important, because it permits the validity of the methods used to be assessed. The question is: have the empirical facts presented been established by validated methods? Whether the purported truthful conclusions are credible, in that they are reasoned using logical principles, is also of paramount importance in evaluating scientific reasoning. The problem of establishing credible evidence (empirical facts) is independent, though linked, to the problem of evaluating the overall truth of a scientific claim.

Scientific laws have the form of general statements, but they are not merely simple generalizations from experience, such as, a claim that "all X are A because some X have been proved to be A." Science is not just an account of the conditions under which observed phenomena occur. Rather, it must include an explanation of the means by which specific effects in certain conditions are brought about. Furthermore, science involves entertaining hypotheses for the purpose of explaining data sets from which the hypotheses are either proved, disproved, or unproved truth or falsity. Confirmation that the hypotheses really are true may be achieved by demonstrating that under similar circumstances the hypothesis is independently proved from a similar but different data set. Scientific knowledge is thus advanced one hypothesis at a time.

In establishing new knowledge, science purports to establish new laws or theories that apply without restriction. Scientific reasoning, roughly characterized, is the way in which knowledge from observed fact may be soundly extended to the unobserved. Any experimental design (intended to establish empirical facts) already involves some formulation of a specific problem, which entails the use of some concepts. An aim of a philosophical analysis is to make the underlying assumptions explicit to clarify and analyze what is implied in using them and to determine whether such concepts fall into a kind of system capable of providing true knowledge (i.e., theories that apply without restriction to space and time). Table 2.1 illustrates the steps in the scientific method.

How scientists arrive at their beliefs (how they reason) is independent from whether those beliefs actually constitute knowledge. The relationship between evidence and conclusions is such that conclusions are logical inferences from the evidence (or premises). From a logical standpoint, the truth-value of the inference depends in part on whether it is justifiable on the grounds it purports to be. Whether one has confidence in the correctness of a certain type of inference is independent of

Table 2.1
Scientific Method

Step	Description
Step 1	Identify a puzzling fact (a sample).
Step 2	Define the problem so that steps to a solution are clear and logically laid out.
Step 3	Construct a hypothesis to account for the as-yet-unexplained fact or observation.
Step 4	Infer a testable consequence from your hypothesis.
Step 5	Test the hypotheses by examining or testing possible consequences.
Step 6	To accomplish step 5, obtain data for the purpose of establishing evidence.
Step 7	Evaluate collected data to determine whether data constitute evidence. Evaluate using logical reasoning, statistical analyses, or experiments (such as surveys, interviews, modeling, graphs) against possible solutions.
Step 8	Conclude from the test results a probable explanation (solution to the problem or failure to obtain a solution, which itself may offer valuable information).
Step 9	Evaluate the chosen proposed solution (or lack thereof) for credibility or truth-value. Two evaluative methods are (1) falsification, trying to prove that the hypothesis is false rather than true; and (2) verification, trying to prove that the hypothesis is true (gathering evidence in support and refutation of the hypothesis).

whether the inference is justified on the basis on which it offered. Likewise, if one uses a logically incorrect method of inferring one fact from others, the facts upon which the inference is based would not constitute evidence for the particular conclusion drawn. Hence, analyzing scientific reasoning involves evaluating the way in which a conclusion (knowledge) from observed fact may be *soundly* extended.

This characterization, however, is simplified. While this description is generally accurate, it is not inclusive of the many varied systematic methodologies that currently comprise science. Furthermore, terms like "proof," "truth," and "fact" in relation to the practice and outcomes of scientific investigations are appropriate for discussions about logical argumentation; however, although they still are used loosely by scientists, current understandings don't include proof and truth. Rather, hypotheses can be disproved by the existence of counterexamples or data that counter a hypothesis. Scientific laws and theories are not, then, absolute immutable truths, and observations are not facts but rather are demonstrations that hold under X conditions at time Y. One of the most fundamental tenets of science is that what might be called a fact can be, and often is, disproved.

Consider the following inductive argument. Most drivers over age 80 are insurance risks. Does it necessarily follow from this fact that any particular elderly individual is an insurance risk? Of course not. Yet the insurance industry (for actuarial reasons) may be correct in assuming that the elderly are insurance risks, and your 80-year-old grandpa is probably not as safe a driver as you. If you rebut that your grandpa is an excellent driver, your refutation might consist of some of the following claims: He's never had an accident. His vision is 20/20. He has no physical or mental impairments, he exercises 20 minutes daily, and so on. However, the form of these rebuttals is the same as the original argument and does not undercut the fact that your 80-year-old grandpa is probably not as safe a driver as you. A refutation of the argument can only be of the same form; the refutation no more follows from the premises than the original argument.

Some statements are not conclusively falsifiable or verifiable by observation or empirical testing. If it is reasonable to accept such statements as true, then it is reasonable to accept those statements as conclusions of arguments in which the premises fail to logically entail the conclusions. Such arguments are not necessarily truth-preserving or valid deductive arguments, and these traditionally are referred to as inductive arguments. In an inductive argument, the premises are evidence for the conclusion or hypothesis. A good inductive argument is one whose premises, if true, establish a conclusion as being more likely to be true than competing conclusions. An inductive argument's conclusion is never proved absolutely. Rather, we accept it with a degree of probability.

We evaluate the soundness of such arguments on the grounds that evidence must consist of true statements. If the statements are true, then it is reasonable to accept the inferred hypothesis as true. There are various ways in which inductive arguments can appear to be true and reasonable but, in fact, are not. The strength of their argument lies not in the truth evidence contained in sentences but rather in appeals to various extraneous rationales, referred to as fallacies, such as those in Table 2.2.

In science, standards exist to enable us to measure and classify, or describe, a physical phenomenon. These standards, like a meter as a standard of length or amount of foreign particles in a chemical compound as a standard of chemical purity, are unchanging and objectively identified. In science, when we can't agree on a hypothesis or theory, it may be that we can't agree on what the question is and, hence, on what the accurate unit of measurement should be. Science, in other words, is descriptive, whereas ethics is prescriptive. Furthermore, a branch of ethics, called descriptive relativism, says that basic ethical beliefs of different people are societies are different and even conflicting. While different cultures or communities may hold different ethical beliefs (e.g., one culture believes that children should put their parents to death before they get old, whereas we in the United States do not believe such), the difference is factual, or descriptive, not ethical. In this case, our ethics, and that of the other culture *both* rest on the precept that children should do what is best for their parents.

Ethics uses standards, too, but they are not immutable or objectively verified. Ethical standards are rules of choice, decision, action, inaction, obligation, duty, and responsibility. In this sense, ethics is normative. Its standards are norms (or prescriptions) that guide action and very much take into account psychological criteria, such as ability, need, desire, intention, motivation, and the like. Its principles are prescriptive. They tell us how we ought to behave or what we should do, and why it is right. Such considerations have no place in science. There aren't any *oughts* in science.

Ethical standards differ from scientific standards. Ethical principles differ from scientific laws. Importantly, scientific standards and laws differ in their logical relation to one another from ethical standards and principles. The application of scientific standards determines the correctness or incorrectness of our judgments. For example, if a book measures 8 inches in length, then we must judge that the book is truly 8 inches long. Though any of us is able to measure incorrectly and come up with a

Table 2.2
Fallacies That Weaken Arguments

Fallacy	Description
Argument from force	Appeals to force; the proposition is said to be true only because someone in a position of power decrees it.
Ad hominem attack	When an author attempts to refute a position by attacking the person rather than the claims within the argument.
Argument from ignorance	When a proposition is claimed to be true only because it hasn't been proved false, or the proposition is claimed to be false only because it hasn't been proved true. This error occurs frequently in situations in which little evidence exists to confirm or refute a position.
Appeal to pity	An emotional maneuver appealing to the audience's sympathy to persuade them to accept a position.
Appeal to authority	Similar to the appeal to force but differs in that the appeal is to an authority in matters outside his or her realm of expertise.
Hasty generalization	When the claimant cites an atypical example to make a generalized point.
Argument ad populum	When appealing to a presumed popular belief or convention to win assent to a conclusion.
Begging the question	When the claimant assumes what the argument is trying to prove; this is a common error.
Complex question	When the issue or question posed really involves two or more independent questions where the answer to the first question is presumed to be yes.
Equivocation	When an argument turns on a crucial shift in the meaning of a significant word or phrase. For example, the word *desirable* can be used descriptively (that something is desired by people) or prescriptively (that people ought to desire it).
Red herring	Side-tracking the argument from the issue under consideration to a completely different issue. This tends to occur when the real or central issue is problematic and controversial, but a side issue (red herring) is easily accepted. By confusing the red herring with the central issue, the illusion is created that an assent to the red herring is really an assent to the central issue.

(*Continued*)

Table 2.2 (*Continued*)

Straw man	When a simplistic (irrelevant) argument is substituted for the central argument.
Fallacy of the consequent	An invalid argument form where if A is true then B. B is true, then A. Even if the two premises are true, the conclusion may be false.[a] (Related to the naturalistic fallacy discussed above.)

[a]For example, it is true that if Napoleon was German, then he was European (because all Germans are Europeans). But it is false and cannot be deduced that Napoleon was German from the premises that (1) all Germans are Europeans and (2) Napoleon was a European.

number of different measurements (a foot, 6 inches, 7.3 inches, etc.), there is one and only one correct measurement (standard) of the book's length. Ethical standards are modifiable as opposed to scientific standards, which are immutable. For example, while stealing is agreed upon as being wrong ethically (i.e., the ethical standard that stealing is wrong), exceptions to this rule exist that justify, or even prescribe, stealing. In other words, there are occasions in which stealing would be considered a good and not a bad thing to do. In situations in which we would agree that it is *not* wrong to steal, we might justify our position on the basis of modifying moral intuitions or higher moral principles.

The main purpose of ethical theory is to provide coherence and consistency to ethical decision making (determining the right thing to do) and sound rationale for judgments (such as why "X is the wrong thing to do," or why "one is obligated to perform act Z"). In essence, the utility of an ethical theory is its ability to guide and explain particular ethical decisions (or moral decisions about a specific situation). Applying an ethical theory to a specific question of conduct enables one to examine how one's competing values logically relate to one another and to overall goals of moral conduct. Both ethical and scientific reasoning can involve generalizing from the particular to the general, but ethical reasoning—unlike scientific reasoning—is largely analogous and not inductive. Generalizing in ethics involves making arguments that explain why a situation is like another situation, rather than inductive reasoning, which propounds encompassing rules that can be proved factually true. More modern theories of ethics, such as the ethics of care or virtue ethics, however, involve reasoning that is less top-down, from premises to logical conclusion. The ethics of care offers a normative theory of what makes actions right or wrong that emphasizes the importance of relationships. Virtue ethics, another more contemporary theory, emphasizes virtue or moral character, as opposed to deontological theories based on notions of duty or the consequences of actions.

Bioethics and Genetics Ethics

Bioethics is an applied philosophical discipline that deals with ethical issues arising in biological research and its applications. Clinical (bio)-ethics is a practical discipline that provides a structured approach to assist health care professionals in resolving ethical issues in clinical medicine.

Genetic ethics can be viewed as a subdiscipline of bioethics in that it deals with ethical issues arising from genetic research and its applications. Unlike the discipline of ethics or moral philosophy, however, genetic ethics is not only practical in focus but also interdisciplinary, because it involves social and personal dimensions of an issue as well as legal, social policy, religious, and cultural dimensions. The practical focus of bioethics, however, makes it no less critical or significant than theoretical analysis of concepts of morality.

Ethical Analysis: Methods, Mores, and Theories

An ethical theory gives us a common framework to follow when approaching and reasoning through moral conflicts. The history of moral philosophy, from the time of Plato and Aristotle up through the present, offers different theories for determining the right thing to do, and in particular for resolving moral conflicts. It is clearly beyond the confines of this book to discuss the entire history of moral philosophy and ethical theory. However, it is relevant to introduce some of the major theories that are invoked as methods to engage in ethical analysis and conflict resolution. Each theory has its own limitations that make it a useful analytic tool, but only up to a point. It is then up to each of us in our wisdom and capacity for reasoning to make modifications that offer the best solution.

Theory of Consequentialism

The theory of consequentialism holds that what is ethical is that specific consequence that creates the greatest good (or most advantages over disadvantages) for the greatest number of people. This method focuses on the results (or consequences) of possible actions. A benefit of this method is that it seeks to address the impact of decision making. Its real limitation, however, is its inability to provide a basis for evaluating one outcome over another.

Theory of Deontology

Deontology, or deontological ethics, is another method whose name derives from the Greek word *deon*, which means "duty." This theory says that the right thing to do is that which fulfills one's moral obligations or duties. The ten commandments represent a ready example of deontological ethics. A major benefit of this theory is that it guides one away from allowing self-interest to override obligations to others. Its limitation, though, is its de-emphasis on the impact or consequences of decisions. In other words, by focusing solely on the importance of one's duty, one can easily miss other significant aspects of the ethical dilemma.

Theories of Rights

Theories of rights constitute a different ethical framework. These theories seek to identify which specific individuals and social rights are at stake. Of paramount importance is the moral claim of individuals in any particular situation. One strength of these theories is that they focus on the moral centrality of the individual in a situation, that is, whatever she or he is entitled to have. A weakness is the inability of such theories to provide a basis for evaluating and resolving competing rights between individuals. Ethical issues involving pregnant women and their fetuses are often focused as rights issues, where the rights of the woman compete against the rights of the fetus.

Theory of Intuitionism

Intuitionism is arguably the preferred theory of the less reflective. Theories of intuitionism hold that ethical dilemmas can be resolved by appealing to one's intuition about what's right, where the strongest intuition represents the right thing to do. In other words, X is right because "I just know it is" or "I feel exceedingly strongly that it is right." Clearly, such theories can justify the strength of our convictions, but they obviously provide little if any way to judge the merits of one intuition over another or any standard by which to judge the consistency of outcomes over time. Intuitions propelling actions in situations 1, 2, and 3 might be similar, but the outcomes of acting on that intuition are distinctly different. People may have wildly different intuitions about what the right course of action is in these three situations, and we surely would not want to base the correctness of a particular course of conduct on the strength—or fervor—of an intuition.

Theories of Justice

Theories of justice espouse that principles of equality are integral to any normative theory of obligation. Theories of distributive justice, for example, define justice in terms of the distribution of good and evil and the relative comparable treatments of individuals. Theories of retributive justice deal with rules for fairly doling out appropriate punishment. The paradigm of justice is that where two similar individuals are in similar circumstances, they ought to be treated similarly; one of them should not to be treated worse or better than the other. However, these considerations could be outweighed by other considerations. Different moral philosophers, most notably John Rawls (1921–2002), have proposed different rules for determining just treatment. Some have proposed rules that say treating people justly is to treat them according to their "just deserts" or merits. Others have proposed different rules for just or fair treatment as treating people according to needs, their abilities, or both. Accordingly, various criteria for determining what constitutes merit, need, or ability have been proposed. For example, Aristotle defined merit as virtue and justice as the distribution of this good, which, because of its moral correctness, results in happiness. Other criteria have been argued for, including, but not limited to, contribution to social welfare, intelligence, bloodline, skin color, social rank, or wealth. A second example is the equalitarian characteristic of modern democratic societies. A third derives from the Marxist dictum, "From each according to his ability, to each according to his needs." In essence, the relevance of theories of justice is their ability to fairly settle disputes involving competing interests, rights, or obligations. As a practical matter, each of us for the most part is not given an equal chance to achieve all of the virtue we are capable of achieving in our lifetimes. Justice, in particular social justice, attempts to equalize chances for all. Justice, in other words, asks us to do something about cases of special need. For example, we ought to give people with disabilities special attention (actions), because it is only with this special attention that they may have close to an equal chance of enjoying a good life.

Ethical Altruism

Ethical altruism, unlike other moral theories, bases its obligation on the good or bad produced for others, that is, for persons other than the agent doing the act. Some theorists argue that even altruism is self-interest and not other-interest based, because we derive pleasure and satisfaction from helping others. In other words, the pleasure we achieve in helping others motivates us to act altruistically and this motivation and its accompanying feeling are not distinguishable from altruistic acts. Rather, they constitute the basis of such acts and further provide the reason why we behave altruistically. Others retort that, of course, we obtain pleasure from such acts of kindness, but we do not perform such acts because we expect to, or necessarily need to, receive satisfaction. The ethics of altruism is somewhat contentious in philosophical circles, because many argue that our first and most basic obligation is to ourselves and our own prosperity. It is not prudent, the argument goes, to act in ways that undermine this primary obligation.

Contextual Theories

Contemporary normative theories of ethics were developed in the second half of the twentieth century by feminists. These theories emphasize the importance of contextual factors in determining ethical defensibility. To illustrate the contrast, consider the following example. A young child is struggling to stay afloat in a lake. The utilitarian argues that jumping in to rescue the drowning child is the right thing to do because doing so maximizes social benefit. In other words, it is in society's best interest (the interests of the majority of people) to rescue the child. This interest might be based on both the benefits and joys the child gives to others as well as the child's potential to benefit society, either through inventions, advancements, and the like. The deontologist argues that rescuing the child is the right thing to do because as human beings we have an inherent duty to preserve life and help one another, and perhaps more important, we are bound by the fundamental duty to "act onto others as we would have them act onto us." A virtue ethicist defends rescuing the child as the right thing to do not because it maximizes social good or because we are bound by a duty to do so, but rather because it is the benevolent or charitable thing to do, and that is what makes it right.

Ethical Concepts and Principles

Ethical analysis includes some basic concepts, the most relevant of which are autonomy, beneficence, justice, and informed consent. Autonomy means the freedom of an individual to determine a course of action for him or herself, and it is a hallowed concept in the United

States. As such, it celebrates a strong sense of individual entitlement and responsibility. In other cultures, such as Native American, Japanese, and Australian Aboriginal cultures, the primacy of the individual, and their wants, are subsumed within a sense of what's best for the community.

"Nonmaleficence" is a term that means "do no harm." It is the most important principle encoded in the Hippocratic Oath of medicine: "As to disease make a habit of two things, to help or at least to do no harms." Beneficence is the flip side of nonmaleficence. "Beneficence" refers to our duty as humans to care for others and to act in ways that benefit them, when we can do so without putting ourselves in harm's way.

The principle of justice or equality refers to the fair and equitable distribution of resources (burdens and benefits) to society or any subgroup. Various theories of justice exist, each of which provide a basis for determining what constitutes a fair distribution. For example, comparative justice claims that what's fair for one person to receive is determined by balancing the claims of all the other people competing for the same or different things. What's fair for one person to receive is determined by his or her need relative to the other members of the group. For example, if two people are competing for a heart transplant, it is fairer for the one who is more ill and who would die sooner without a heart transplant to receive the first available heart.

Informed consent is a bioethical concept that emerged from unethical medical research practices in Nazi Germany. The Nuremberg Code of Ethics was developed during the Military Tribunals from 1946 to 1949, and stressed informed consent as a standard medical ethic. However, the roots of the concept of informed consent and ethical medical research lie in earlier U.S. history. From 1932 to 1972, the U.S. Public Health Service conducted medical experiments on 399 illiterate and desperately poor African-American male sharecroppers who were in the late stages of syphilis. The Tuskegee experiments involved telling the men that they were being treated for "bad blood," although researchers had no intention of curing them of the disease. The data were collected on autopsies of the men. The true nature of the research had to be kept secret from the men to ensure their cooperation. Furthermore, as one of the physician researchers said, "As I see it, we have no further interest in these patients until they die." The research flagrantly violated the Hippocratic Oath, in which the physician is "to do no harm." Another example is the U.S. government's secret plutonium testing on thousands of citizens from the 1940s to the early 1970s in the guise of preparedness for the Cold War. These citizens were the innocent victims of more than 4,000 secret classified radiation experiments conducted by the Atomic Energy Commission and other federal agencies, including the Departments of Defense, Health, and Education and Welfare; the Public Health, now the Centers for Disease Control and Prevention; the Veterans Administration; the Central Intelligence Agency; and National Aeronautics and Space Administration, after which Congress strengthened regulations.

Informed consent has become a cornerstone of both medical research and medical care. It is now a legislated practice in every state. In other words, it is not only immoral, but illegal, to conduct research on human subjects unless they are informed beforehand of all the known potential risks and benefits of participating. To conduct research or provide care without a valid informed consent is immoral and illegal, and it constitutes battery. Obtaining valid informed consent requires that potential subjects are able to understand these conditions. Research subjects and patients who undergo various medical procedures or therapies (e.g., chemotherapy) must be informed ahead of the study or procedure about every known and potential benefit and risk of the proposed intervention. If a patient is unable to understand these conditions, a surrogate decision maker (parent or guardian) must provide the informed consent. There are, however, many medical interventions and therapies that we are not required to consent to, such as having an x-ray taken, taking medicines (though the prescribing physician or pharmacist may tell you what the known risk of side effects is in addition to benefits you can expect), having a cast put on, and having a blood or urine test.

These four principles (autonomy, beneficence, justice, and informed consent) provide an underlying structure from which ethical decision making and analysis occurs in genetic research and its application in clinical medicine.

CHAPTER 3

The Use and Abuse of Genetic Information: Genetic Privacy and Genetic Discrimination

British researchers reported in 2004 that although doctors think the results of genetic tests should be available to insurers, it is unlikely that insurers will misuse test results, even though it could be in their financial interest to do so.[1] Do you agree? In other words, can you imagine third parties (such as insurers, employers, or educational institutions) obtaining private genetic information without the authority to do so (without prior consent)? Can you imagine these institutions using such to maximize their own self-interest and, in so doing, adversely affect the people who's genetic information they obtained? Do you think genetic discrimination exists, or that it soon will be a reality? If so, do you believe it will inevitably go unnoticed or unpunished, because such acts are typically covert and difficult to prove in court? If you haven't given these, and similar questions about the appropriate use and misuse of genetic information much consideration, rest assured that others have. At root is the ethical question of when (if ever) biology should provide a basis for differential treatment. The stakes involved are high for individuals, their families, and ultimately society itself.

The question of what constitutes fair treatment when biologically different individuals compete for the same benefits and entitlements is not new. Genetic information just adds a new, albeit unique, biological basis. Separate, and no less significant, matters, are questions about who is entitled to which societal opportunities and benefits, particularly when individual rights are pitted against one another or against group or institutional rights. Resolving ethical issues about potential nonuniformity among different nations' decisions about the conditions of appropriate and inappropriate use are the tasks of many international committee meetings. Countries around the world agree that genetic research ought to progress, for it promises substantial human benefit in various areas (medicine, agriculture, environmental protection, etc.), but they differ in how they define distributing access fairly, as well as in how to manage a host of other ethical concerns.

The promised benefits of genetic advances may be achievable only at the cost of other actual, or potential, burdens to individuals and their blood relatives. The reality is that optimal benefits and negligible harms for all is a myth because of the many and varied competing stakeholder interests. Trade-offs are inevitable. Harm to some is inevitable. Fairly balancing the interests of current stakeholders, and the anticipated interests of future generations, is a normative challenge. This challenge, in part, involves determining who is permitted to access what genetic information under what circumstances, and what constitutes appropriate and inappropriate use. While appropriate uses for genetic information are highly contentious, depending on the respective interests of stakeholders, inappropriate uses are perhaps more clearly defined by situations resulting from obvious misconceptions about the truth-value of cutting-edge science or the willful neglect of the truth.

Notably, at the inception of the Human Genome Project, there was international concern that the genetic information could create a social underclass of "genetic" undesirables. This prompted ethical, legal, and social policy research, national policy statements, and international policy recommendations by United Nations Educational, Scientific, and Cultural Organization (UNESCO) and several other influential international organizations. These actions were intended to preempt social injustices resulting from inappropriate access or misuse of private genetic information. In the United States, state and federal legislators have considered these and related possibilities. In efforts to prevent exclusions based on beliefs about the meaning of genetic information, states have enacted antidiscrimination laws, and Congress has passed legislation intended to protect genetic privacy and considered several bills outlawing genetic discrimination. Nearly all of the 50 states now have some form of prohibition legislation, though the definitions of genetic information and the scope of prohibition vary considerably. On the federal level, President Clinton, on February 8, 2000,[2] recalling that Justices Warren and Brandeis who, well over 100 years ago foresaw the urgent need to safeguard individual privacy from the implementation of technological advances that could pose serious threats to this fundament liberty, issued an executive order banning genetic discrimination in federal employment. In that order, Clinton proclaimed that genetic advances pose a profound threat to "civilization's most valued liberty: privacy." In sum, substantial resources have been spent to ensure that learning one's genetic makeup, if one chooses to do so, will be far more beneficial than harmful.

Yet, despite the existence of legal protections, concern remains as to whether they will succeed in preventing abuses, particularly as these laws have been largely untested. Moreover, significant concerns exist about the potential for adverse impacts in areas not proscribed by law. What (if any) harm might affect individuals, their families, communities, social institutions, and cultures of peoples around the globe remains unknown. As discussed in chapter 1, history reveals a certain human penchant for excluding individuals on the basis of differences (real or perceived) from the desired norm. It is largely owing to this history that many suspect that genetic variation associated with undesirable characteristics will provide a new venue for ostracization, stigmatization, and the creation of an undesirable genetic underclass.

Whether in fact personal genetic information is inappropriately (even illegally) obtained and whether its use in turn (deliberately or not) results in adverse consequences (discrimination, privacy violations, confidentiality breaches, coerced testing, and the like) is a hotly debated matter. Were the potential for these types of abuses minimal, then it is highly unlikely that the state and federal prohibitive measures would have implemented. This is to say that the potential for misuse is great. Legal cases alleging misuse have been documented and researchers have reported numerous types of exclusions to social benefits and entitlements, based on genetic information. Others disclaim the veracity of a problem on grounds that cases are anecdotal and unproven. Nonetheless a majority of states have legislation in place to protect individuals' genetic privacy and prohibit genetic discrimination, though it is important to note that these laws are not consistent with one another and have been infrequently tested.

Importantly, third parties do and will continue to have a financial stake in a genetic test and so the potential for misuse exists. Companies with such an interest might, for example, test individuals without their knowledge or consent and nonetheless act on test results. In fact, one such case occurred recently. The Burlington Northern Santa Fe Railroad tested 35 employees[3] who had sought disability compensation for what they claimed was job-related carpal tunnel syndrome, for hereditable neuropathy in efforts to disprove that the condition was work-induced, hoping instead to prove that it was a preexisting genetic condition. If the workers tested positive, the employer could claim that the symptomatology was consistent with clinical expression of mutations associated with hereditary neuropathy and, hence, did not qualify for workman's compensation. Also important to the case was the fact that the employees did not know that they were being tested. Blood they were obligated to give during their routine employment physical was tested without their authorization/consent. Moreover, it was only due to sheer accident that that testing came to light. The incident first came to the attention of one employee. Soon thereafter the others who had been tested learned their similar fate and sued the company. Genetic test results were uniformly negative, meaning no employee tested positive for hereditable neuropathy. What the company failed to know, or willfully neglected, is that hereditary neuropathies occur very rarely in the population, making it highly unlikely that a group of employees all had the condition. Furthermore, carpal tunnel syndrome is one of many symptomatologies that can develop when the condition is present.

Independent of concerns that individuals will be tested for genetic disorders without their prior consent and/or discriminated on the basis of symptomatologies of disease, is concern that the identification of disease-related genotypes, particularly in the absence of clinical manifestation, could lead to exclusions (discrimination) based on genotypes either known or, importantly, presumed. Such adverse differential treatment on the basis of genetic status is referred to as genetic discrimination.[4] To illustrate how potent such discrimination can be to any individual, given that each of us harbor five to seven mutations capable of causing disease, consider the following story. To protect privacy the names have been changed.

Kay's Story

The adverse effects of ostracization can run deep, but arguably none run deeper than those based on genotype, as they pierce the core of who we are as individuals. While most teens view their lives with limitless possibility, Kay knew that her family's history of Huntington disease could deprive her of all her dreams. Having directly witnessed the disease ravaging the lives of her father and grandfather, and convinced that she too would succumb to Huntington disease, Kay carefully planned her life certain not only that she had inherited the gene but that because of that "flaw" she was not worthy of the opportunities that are open to all. She dated only men with disabilities and believed she was not entitled to have or adopt children.[5] She secured a government job that afforded her solid benefits that would cover her illness. In her late 30s, shortly after DNA-based testing for Huntington disease gene testing became available, Kay underwent testing and learned that she did not carry the mutated gene. She would not develop Huntington disease. The basis on which her life choices had been made was suddenly false. The self-loathing that had fueled her goals gave way to survivor's guilt. In the end, however, she was able to remove the shackles of despair and go on to lead a happier life. Her experience illustrates both the devastation that foreknowledge of a lethal genetic disease can have as well as the potent influence our perceptions about our genotypes carry regardless of whether based on fact or not.

Genetic Privacy and Genetic Discrimination

Although the human genome has been fully mapped and sequenced, and our fund of genetic knowledge is growing, vast mysteries remain about how and why human genetic composition manifests with such variety. Though the era of full genome scanning replete with chip-based result interpretation does not yet exist, many anticipate a growing reliance on genetic information to provide better quality health care. The potential medical value of genetic information, as discussed earlier, lies in its promise to improve detection, treatment, and prevention of a nearly limitless number of diseases. As genomic medicine becomes a reality, multiplex testing (i.e., testing for many genes and gene variants at once) creates potential competing utilities of predictive information. Insurers clearly can profit from having predictive information about their existing, and potential, clients for more precise underwriting. Theoretically, their underwriting could be tailored to more precisely reflect individual risk, so perhaps they would charge a higher premium or refuse to cover those deemed too great a risk. The actual value of predictive information may be lower for individuals, particularly if the therapeutic value of such information is uncertain, or the indicated treatments unaffordable.

Despite the immutability of our individual genomes, what we understand today about the function of specific genes in the future may be quite different. Variants associated with particular protective effects could become known to confer other adverse effects, or vice versa. For example, the mutation in the gene for hemoglobin that causes sickle cell anemia also confers a partial protective affect against malaria. Given that what we now know about gene mutations, common variants, and their biochemical products will likely change in the future, this unknown knowledge has important implications for preserving genetic privacy. What one consents to knowing at one time may result in knowing other things that, if one had the foreknowledge of that possibility, he or she may not want to know. For example, if you were to have a full genome scan today, you may learn that you have a gene variant that indicates you should not be given codeine for pain. The medicine is not effective because of your genetic makeup. Twenty years from now, however, this, same genetic variant may indicate not only the need to avoid codeine, but also that you have an increased risk for a serious and highly degenerative neurological disorder. So what you learn from a test today might come to indicate susceptibility to a wide range of other disease risks, traits, or behavioral susceptibilities that weren't known at the time when your sample was tested and, importantly, for which your consent was not possible or remotely adequate. In other words, while our knowledge of human genetics is expanding at a rate akin to the speed of light, our understanding of what the information means, and how best to use it, constitute formidable ethical concerns.

What Is Genetic Information?

Genetic information is defined as the data about an individual's genetic makeup, that is, the unique sequence of DNA (deoxyribonucleic acid). It may be inferred from an individual's family history, from various routine chemical tests that have genetic implication (e.g., elevated hemoglobin A1c seen in uncontrolled diabetes), and from direct DNA-based or chromosomal tests. Some genetic information is inherently public and not private. For example, physical traits, such as skin color, hair color, height, and skeletal structure are all expressions of one's genes. Other genetic information is not outwardly obvious to others and is considered by most to be intimately private information, such as inherited susceptibilities to specific diseases.

Your personal genetic information might be valuable to you. Knowing whether you've inherited risks for cardiac and neurologic disease might well influence your ability to prevent or successfully treat disease, as well as influence your perception of your life opportunities. This could include your decisions about your occupation, whether to marry, whether to have children, and so on, as exemplified by Kay's story.

Genetic information is predictive, albeit to a probabilistic degree. Test results for single gene autosomal dominant disorders (i.e., Huntington disease [HD]) can show whether an individual has the expansion in the triple nucleotide region of a gene that is associated with that disorder. Despite the test's **sensitivity**, it cannot tell an individual when they will develop the disease. For diseases with variable expression, a test result predictive of disease does not predict the severity of the disease. Disease expression is determined by the interaction of numerous environmental factors and possibly other biological factors, not simply a change in a DNA sequence. Furthermore, though a genetic test is highly predictive, the risk a result confers is best understood in a context of all of life's uncertainties. In other words, alternative possibilities (like death in a motor vehicle crash) could well occur before any disease developed. For example, a woman with a *BRCA* mutation has a significantly increased risk of developing breast cancer. However, she could nonetheless develop breast cancer from causes unrelated to the *BRCA* mutation, such as other genetic, gene-environmental, or environmental contributions.

Learning one's genotype can be immensely desirable, as well as powerful. It's not hard to imagine that someone would want to have this information for sheer curiosity as well as to prevent disease or optimize reproductive choice. It is equally not difficult to imagine that someone else wouldn't want to have this information, preferring ignorance to any possible adverse psychological or medical effects. Parents might want their children tested to take preventive measures if indicated to ward off future or late onset disease, but it's not difficult to imagine that the child, once becoming an adult, may regret having that information, particularly since he/she did not consent to have the testing done. Furthermore, the reality of testing might well prove to be distinctly different from what one imagines or anticipates. In practice, a prediction of how a test result might affect the individual is not likely to be 100 percent accurate. It's one thing to want to undergo testing and even be certified as an appropriate candidate. It may be quite an emotional jolt if you think you want to know your genetic profile, but upon getting your results, you regret having that knowledge.

The vast majority of human genetic information will not be about single-gene autosomal dominant disorders, because these are comparatively rare and therefore affect a relatively small proportion of the population. Rather, the majority of human genetic information generated will consist of gene variants associated with increased or decreased risks for diseases or efficiencies in metabolizing different agents, such as nutrients, toxins, or medicines. None of the information obtained is 100 percent certain, though probability factors can come close to that percentage, as is occurring now in much of forensic testing, including DNA testing to confirm identity in the 911 disaster or in alleged rape cases. Few, if any, tests can determine, every known mutation. For example, the test for cystic fibrosis (CF) carriers only screens for a few hundred of the most common mutations from more than 900 of those known to be associated with CF. Hence, a negative test does not give a 100 percent assurance that an individual is not a carrier of a CF mutation. Obtaining your genetic information, then, can tell you a lot, but your biology is by no means your destiny. We are, and always will be, far more than our genomes.

Moreover, equally important to analyzing the possible impact of actual genetic information is the possible impact of presumed or perceived genetic information. The complex nature of genetic testing coupled with the difficulty in correctly interpreting the meaning of results, due in part to the comparatively small number of trained individuals, increases the possibility of misconstruing the meaning of a result. In other words, people may misunderstand the meaning of a genotype, or specific piece of genetic information, and come to erroneous beliefs based on this falsehood. Misperception and misinformation about the meaning of genetic information, as stated earlier, also spurs concern about the potential for genetic discrimination.

The So-Called Argument for Genetic Exceptionalism

Many have debated whether genetic information is unique, or at least distinctly different, from medical information. That is, is it so exceptional that it requires special protection? Medical information includes records of past illnesses and injuries; personal health-related habits, such as diet and exercise; family medical history; and health status checks, such as weight, height, blood pressure, cholesterol, and heart rate. Much of an individual's medical information, as opposed to their genetic information, is changeable. One can gain and lose weight, blood pressure can go up or down, and so on. Genetic information, by contrast, is immutable. It will not change over time as a result of your health habits; however, the *expression* of your genes might change as a result of lifestyle and dietary changes. Some have tried to distinguish medical from genetic information on the grounds that genetic information is generated by genetic tests. However, biochemical tests can diagnose a genetic condition. They can, for example, identify enzymatic dysfunction indicative of a genetically based metabolic condition or an elevated serum ferritin and transferring saturation level that demonstrates the presence of hereditary hemochromatosis. In addition to being immutable, DNA-based testing, unlike other types of medical testing, is informative not only of the patient being tested, but their blood relatives as well.

What Is Genetic Discrimination?

The term *genetic discrimination* is often blurred to mean exclusions based on genetic information in the presence or absence of phenotypic considerations. Originally, many used this term to refer to genotype-based discrimination, as distinct from phenotype-based discrimination, that is, the presence of symptomatology, because the latter arguably includes race, gender, age, and disability-based discrimination (the more traditional reasons for discrimination). Phenotypic-based discrimination extends solely to that individual, whereas genotypic discrimination is broader and conceivably may extend to any blood relative about whom the genetic information is informative. For our purposes here, genetic discrimination is defined as differential treatment based on an identified or a presumed genotype. In other words, someone might presume your genotype by inferring it from either a relative's genetic test result or a known family history.

Present concerns about the possibility of genetic discrimination stems from various sources. First are actual reported occurrences. Second is historical evidence that people have used various scientific pretexts to discriminate against one another over time. Third, is the fact that existing legal protection is largely untested, so its effectiveness to protect people is substantially uncertain. Fourth, the Americans with Disabilities Act of 1990 (ADA), does not unequivocally cover genotype-based exclusions. A current debate is whether a genotype, such as predicted high risk of late onset disease, in the absence of symptomatology, constitutes a disability.

Historical Penchant for Exclusion

Although *genetic discrimination* is a recently coined term, exclusions based on biological differences have a long social history. Some of this history is discussed in chapters 1 and 4. What is evident throughout time are the repeated instances in which the application of new science resulted in adverse social consequences. The inherent value of predictability, one could argue, lies in its applicability. However, an important normative question is whether predictive "knowledge" ought to be used to improve society or better humankind's lot. Independent of that fundamental ethical concern are normative questions about feasibility, such as how best to maximize benefit while minimizing harm.

In his book *Heredity in Relation to Eugenics* published in 1911,[6] Charles B. Davenport (a zoologist and one of the early American eugenicists) expounded on the social utility in applying so-called *genetic truisms*:

> Recent great advances in our knowledge of heredity have revolutionized the methods of agriculturists in improving domesticated plants and animals. It was early recognized that this new knowledge would have a far-reaching influence upon certain problems of human society—the problems of the unsocial classes, of immigration, of population, of effectiveness, of health and vigor.... It is a reproach to our intelligence that we as a people proud in other respects of our control of nature, should have to support about half a million insane, feeble-minded, epileptic, blind and deaf, 80,000 prisoners and 100,000 paupers at a cost of over 100 million dollars per year.... But we have become so used to crime, disease and degeneracy that we take them as necessary evils. That they were so in the world's ignorance is granted; that they must remain so is denied.

Another prominent eugenicist, Harry Laughlin, also worked steadily to refuse immigrant Jews and Irishmen entrance into the United States.[7] To the New York State Special Commission on Immigration and Alien Insane he wrote:

> Much of crime in America was perpetuated by recent immigrants and aliens, and that America would be far better off if it restricted future immigration to the white race, and within it established differential numerical quotas to admit individuals only on the basis of desirable racial elements so as to improve the future "seed-stock" of America.[8]

Garland Allen, a social historian with expertise on the American Eugenicist Movement, argues that the movement rose and fell primarily because it served a distinct need of the elites in the United States at that time. When the ruling class had accomplished its purpose of compulsory sterilization laws and societal indoctrination of "the Fitter Family," he argues that the movement ended.[9] While Allen's theory may be correct in explaining the acceptance and rejection of the Eugenics Movement, genetic discrimination has the potential for unintentional self-induced and intrafamily-induced harms.

Later, scientists demonstrated that various phenotypes conformed to Mendelian predictions of inheritance, specifically, ABO blood groups, red-green color blindness, and certain diseases like inborn errors of metabolism. Such discoveries illustrated how human phenotypes could be prevented through preventing matches between certain genotypes. Some, particularly those within the disabled community, claim that phenotypic prevention via genotypic prevention remains a dominant prevention strategy that is all too often uncritically accepted as not only ethically appropriate but *the only* means of prevention. This strategy is false.

The late 1950s' discovery that the human Y chromosome plays a major role in sex determination, and the further elucidation of Kleinfelter syndrome (males with too many X chromosomes [XXY]) and Turner syndrome (females with only a single X chromosome [XO] rather than the normal two [XX]) by karyotyping highlighted the importance of the Y chromosome as being more important to male development than the X chromosome. This new insight arguably spurred curiosity about the possibility that the Y chromosome predisposes an individual to violent behavior. In light of the preponderance of males in penal institutions, research studies were soon under way. As will be discussed in chapter 4, the ethics of such research continues to be contentious. A case in point occurred in the early 1990s. The National Institutes of Health (NIH) had funded the Genetic Factors in Crime conference to convene experts in psychological, physiological, and molecular biology research as well as in criminology and criminal law to evaluate various research approaches, study findings, and societal implications.[10] Controversy heightened to the point that the conference was cancelled. In 1995, the conference resurfaced as a meeting closed to the public that was held at the University of Maryland. Protestors, nonetheless, succeeded in barging in to issue complaints and then were ushered away by police. Research in this area continues today using newer, more sophisticated molecular techniques. While no specific studies have been halted for ethical concerns, the field remains highly controversial.[11]

The early 1950s also saw the Neel and Beet's discovery[12] of how the abnormal sickle cell was inherited, which spurred interest in using this knowledge for disease prevention. Symptoms of sickle cell have been traced back to 1670[13] in one Ghanian family. Although the abnormal sickle cell was first identified and described in 1910 by James Herrick, James Neel in 1951 was the first to distinguish sickle cell disease from sickle cell trait, which he described as a condition in which a small percentage of red blood cells have the sickle shape but behave in all other respects like normal red blood cells. Scientists then identified several substances, such as urea and cyanate, which proved moderately successful in helping patients cope with sickle cell crises, but a cure was nonexistent. Importantly though, it was believed at the time that sickle cell trait was a benign form of sickle cell disease (which it is now known to be incorrect) that in the presence of particular precipitating environmental factors (e.g., high altitudes, infection, underwater swimming, and alcoholic intoxication) could produce symptoms identical to those of disease. Given that symptomatology is either absent or infrequent for the trait, its diagnosis could occur only through specific biochemical tests and not via clinical manifestations. Several techniques, such as the stained blood-smear examination, sickle cell slide preparation, sickle-turbidity tubule test, and hemoglobin electrophoresis, were devised as diagnostics.

At the time, it was estimated that at least 50,000 African Americans had sickle cell disease and as many as two million had the sickle cell trait. As opposed to the community at large, researchers and public health officials recognized sickle cell disease as a sorely unattended public health problem, citing a dearth of research on the disease and the absence of health care planning objectives.[14] Racial bias was cited as a contributing factor to this oversight. From a sociopolitical public health perspective, there appeared to be a need to prevent it from spreading. In the absence of a cure, genotypic prevention by identifying those who were affected, and hence carriers, would permit selection, termination, and early diagnosis.

In 1971, despite the lack of effective treatment and lack of mandated funding for genetic counseling, compulsory sickle cell screening was adopted by many states. By 1973, 10 states had passed laws requiring screening, which affected roughly 40 percent of the African-American population, while four others had passed voluntary screening laws. The imposed mandates were inconsistent both in rationale and targeted subgroups. Some required newborn screening, and others required the screening of preschool children, with school admittance contingent on testing certificates. Other laws required screening inmates of state institutions and testing couples as a condition for marriage licensure. More important, the rationale underlying these mandates often ranged from ill informed to immoral, with newborn testing for carrier status evincing this point.

Furthermore, inasmuch as these laws lacked provisions for genetic counseling or education, the social risks of participation, including but not limited to privacy and confidentiality risks, were not adequately communicated. Lack of confidentiality proved to be a genuine problem.

Some individuals whose trait identity became known incurred discrimination, most notably job discrimination, exclusions from military service, and refusals in the areas of adoption, education, and social stigmatization.[15] Ignorance and misinformation, not only on the part of those undergoing screening, but also on the part of those implementing programs, contributed to the resulting adverse psychosocial consequences. Confusion in understanding the difference between being a carrier of sickle cell trait and actually having sickle cell disease was largely to blame. In addition, some argue that policy makers apparently failed to consider the possible eugenic implications of informing someone of their trait (carrier) status.[16]

In the context of racial tensions of the early 1970s (carryovers from urban riots of the late 1960s), screening programs faced mounting racism charges and growing public opposition that eventually culminated in the passage of the National Sickle Cell Anemia Control Act of 1972. The act mandated the use of federal funds for voluntary screening programs only and thus brought a halt to mandatory laws. Federal funding spurred states to amend their mandatory laws favoring voluntary testing; other states opted to abandon state-controlled testing altogether. Prenatal screening was not mandated largely because there appeared to be little interest in it, especially since sickle cell disease is variable and not ultimately fatal at an early age.

The possibility of adverse consequences and the potential social risks posed to individuals and their families in undergoing screening tests were considered and identified only *after* the implementation of programs. For example, the Odessa Brown Children's Clinic,[17] which tested 1,930 children between July 1, 1971, and October 31, 1972, evaluated its program 14 months after screening began. Considerable confusion about the meaning about the sickle cell trait among the families of those being tested was a primary finding, with respondents (in this case parents, which made up 43 percent of the carrier group and 71 percent of control group) considered sickle cell trait to be a disease. Parents were led to believe that carrier status required frequent medical care and dietary supplements, that any preexisting medical conditions were exacerbated by the presence of the trait, and that curtailing strenuous activities, like sports that would expose the child to diminished oxygen intake, was an important preventive measure. Evaluators found that no organized protocol for genetic counseling existed, which suggested that misinformation on the part of practitioners may have contributed to parents' confusion. Such misinformation is believed to have contributed to the loss of employment and insurance benefits, exclusions from military service, and other psychosocial burdens, such as the *vulnerable child syndrome*. The vulnerable child syndrome refers to parents viewing their children as being different, in a negative sense, and that there must be something wrong with them, in this case, because a physician told them that their child carried the sickle cell trait.

Advocates of screening programs contended that trait carriers ought to be informed of their genetic status to promote fully informed reproductive decision making. The underlying assumption was that fewer carriers would mate and, subsequently, the number of babies with sickle cell disease would decrease; again, this is an example of phenotypic prevention via genotype prevention. George Stamatoyannopoulos, a prominent sickle cell researcher, refuted this assumption in a 1974 study that found that the identification and counseling of heterozygotes did not alter mating behavior.[18]

In 1987, the NIH Consensus[19] recommended universal newborn screening for hemoglobinopathies, including sickle cell disease, based on improved treatments and a desire to avoid any racial bias. Particularly influential to this decision was the fact that racial self-reporting can be misleading as to who needs testing, as well as the discovery that sickle cell disease can occur in persons of Mediterranean descent. Because of varying perceptions of one's ethnic identity, self-reported ethnicity can be an unreliable tool for determining who is at risk and therefore an appropriate test candidate. The many state newborn screening programs existing today were reinstituted on the basis of these NIH recommendations as well as increased federal funding.

Genetic Privacy and Confidentiality of Genetic Information

The *Oxford American College Dictionary*[20] defines *privacy* as "being private, seclusion." That definition seems not to be terribly informative, especially considering the fact that privacy is understood within a cultural context. Different cultures, or even communities within a culture, have particular norms governing what is and what is not private. *Oxford*'s definition of *private* is more helpful: "(1) of or belonging to a particular person or persons, not public, as in private property; (2) not holding public office, a private person and not an official or public performer; (3) not to be made known publicly, confidential." The same dictionary defines confidential as "to be kept secret, entrusted with secrets." Clearly, privacy and confidentiality are intertwined and even complementary. It is worth noting that the terms themselves are contextually defined.

A legal definition of privacy is different from a bioethical definition, and both differ from a definition of confidentiality within medical practices.

Legal Definition of Privacy

While there is no Constitutional right to privacy, legal bases for privacy rights can be found in the First, Third, Fourth, Fifth, and Ninth Amendments. The concept was first considered legally, with regards to deserving a remedy for infractions, when Justices Warren and Brandeis, in their 1890 foundational treatise "The Right to Privacy,"[21] extended the notion of liberty and property to recognize a new right—the right to privacy—as the "right to be left alone."

Bioethics Definition of Privacy

In *Principles of Bioethics*,[22] Beauchamp and Childress define privacy as a "state or condition of physical or informational inaccessibility." According to the authors, privacy does *not* include the right to control access. Someone may have privacy, not because they have willed it, but because others ignore them. A careful distinction between *autonomy* and *privacy* is drawn.

Confidentiality within Medical Practice

Confidentiality in medical practices refers to the obligation of health care providers to not divulge personal medical information. The notion originates from the Hippocratic Oath and applies to individuals in and outside of their connection to their practice. Confidentiality may be ethically breached if great harm could come to a patient—or society—by not disclosing information to intervening parties.

Genetics and Competing Rights to Privacy

The subtle distinctions between notions of privacy, and the power of their respective implications are arguably even more complicated in the context of genetics, where the right to know or not know genetic information implicates more than one person (a person's genotype informative of his/her blood relatives), and because third parties, such as employers, insurers, or the government, may have an interest in obtaining that information as well. Employers, may want to ensure that their health benefits costs are as low as possible, and whether an immediate family member of the employee has a genetic condition, or predisposition may affect their willingness to insure that individual. Employers want a healthy and productive workforce, so for reasons of self-interest they may want to exclude potential employees who are at risk for serious diseases. Similarly, employers may want to know which employees are at high risk for disease if exposed to certain environmental factors and may bar them from working in conditions that pose extreme hazards to maintain a healthy workforce while warding off potential liability suits.

Increasingly, genetic tests to identify hereditable susceptibility to environmental exposures will be available, and it is not improbable that employers and their, employees, would want that information. One such test is for chronic beryllium disease (CBD), which identifies an inherited susceptibility to developing this lung disease if exposed to beryllium. Notably, the disease does not develop without exposure. For affected individuals, exposure to beryllium dust (found in certain workplaces) can result in this chronic lung-scarring disease. Here, the body's immune system attacks and attempts to break down the beryllium, resulting in lung scarring known as granulomas. The scarring causes the lungs to stiffen, making it difficult to breathe and obstructing the transfer of oxygen from the lungs to the blood stream. CBD can develop in people with only brief or low levels of exposure. More recently, a major U.S. beryllium manufacturing company offered CBD susceptibility testing to employment applicants, raising even more complicated ethical issues caused by the limitations of available testing and the predictability of results.[23]

Employers, for altruistic and/or self-interest reasons, may want to exclude employees from working in conditions that pose extreme hazards to their health. Battery factories, for example, have lawfully excluded women of childbearing age from particular jobs that expose them to teratogens, which have been shown to harm pregnancies.[24] Unfortunately, these assembly-line jobs commanded a higher salary. Conflicts between an individual's rights to know and rights to self-determination and an employer's rights to protect workers from potential harms can arise in these situations. For example, consider a brilliant scientist with a hereditable predisposition to an environmentally induced and lethal disease (such as CBD). Consider further that this scientist wants to work in his field of choice, which would expose him to the environmental inducers of the disease he is genetically vulnerable to; however, if permitted to work this field, he could (for the sake of argument, will) make a discovery of enormous benefit to all humankind. If the worker voluntarily agrees to assume all risks associated with this exposure (e.g., he/she agrees to waive all rights to sue the employer), should that individual be permitted to exercise their autonomy and work as they desire, particularly in light of the likelihood that such work will confer substantial benefits to all? This is one of many other difficult ethical questions about how to balance individual rights against social (institutional) rights, involving workplace issues as well as other societal issues.

Questions of privacy arise also in the context of disclosing family history. Typically, patients disclose private family history during clinic visits because it is in their best interest to do so. Though such is the norm, it is typically not the case that family members have provided prior consent to such disclosures. This raises ethical questions about whether patients have the right to disclose private medical information of family members without prior consent The most vivid actual case illustrating ethical dilemmas about consent to disclose in the context of intrafamilial interests is a 1998 clinical research case. In this case, a clinical genetics research subject provided her father's family history information, as per the study protocol, but did not ask her father's prior approval. When the father learned that his private medical information had been disclosed, without his prior consent, he blocked researchers' access to his medical record, arguing that he did not consent to disclosure and that any third-party access without his prior consent amounted to a right to privacy violation. The Office for the Human Research Protections (OHPR) halted the research to clarify this issue and later determined in this case that the father's information could not be accessed without his prior consent. The case illustrates that family history intakes involve the transmission of third-party information that one is not normally required to obtain prior permission to disclose, and that, even among family members, conflict of interest issues can arise that pit one relative's rights to know against another's right to privacy. The question of whether disclosure of family history information requires the consent of every family relative remains uncertain, as does whether consent applies to deceased individuals, and if so, how.

Related to issues of privacy rights are ethical issues related to whether it is morally acceptable if not imperative to breach privacy (confidentiality) in certain instances to achieve a greater good. Such concerns have surfaced in the context of doctor-patient relationships, in which the question arises of whether it is morally acceptable, if not required, to break doctor-patient confidentiality to achieve a primary moral good. The 1976 Tarasoff case,[25] in which the Supreme Court of California ruled that a psychiatrist has a duty to warn individuals about dangerous patients, set a legal precedent for challenging absolute privacy between doctor and patient. Subsequent to that case, and specific to genetic transmissibility of disease and disease risk, the courts have addressed the question of whether doctors have a duty to warn family members about recurrent risk.

In March 1987, the Florida Supreme Court addressed the issue of whether physicians have a duty to warn family members about genetic transmissibility.[26] The case involved a woman treated for medullary thyroid carcinoma whose daughter received the same diagnosis three years later. Alleging that the mother's physician knew, or should have known, that the disease was heritable and therefore was obligated to inform the mother that her children should be tested, the daughter filed a malpractice suit against the mother's physician. Early testing, the daughter argued, would have enabled her to not only take precautionary health measures, but also consider the probability of recurrent risk in reproductive decision making. The court found that the physician *did* have a duty to warn about the genetically transmissibility of the condition and that this duty would have been met by warning the mother that the condition was transferable to her offspring. Normally, only the patient who is in privity with the physician has a cause of action. In this case, however, it was found that there was obvious benefit to certain identified third parties, and in cases in which the physician knew of the existence of these third parties, the physician's duty extended to them as well. A similar case occurred in New Jersey in 1996 when a daughter sued her father's physician's estate when she was diagnosed with adenomatous polyposis coli, 25 years after her father was diagnosed, claiming that the physician should have warned her about her 50 percent risk of developing the condition. The court found that physician did have a duty to warn the daughter directly, even over the father's objections.[27]

These cases illustrate some of the many complicated issues that can arise among family members with competing interests in obtaining or guarding their genetic information. Furthermore, the real possibility exists that **biomarker** information, like other hereditable information, could create new duties and liabilities, particularly when clinical significance is at issue. While physicians arguably have an ethical obligation to understand the genetic disorders that occur with increased frequency in their patient population, they are not obligated to offer genetic tests or warn family members, particularly in the absence of standards of care.

Another ethical dilemma involving competing rights is best illustrated in situations involving genetically identical twins. What ought to happen, for example, when one twin wants to know their genotype and the other does not? Can both a right to know and a right not to know be preserved? How can we enable the rights of one without violating the rights of the other? Which twin's right is primary? If neither takes precedence and both are equal, how ought we balance one twin's right to know with the other's right not to know? Specific factors might impact the balance, such as the existence of a proven medical treatment. Even so, who ought to decide which factors are relevant and which are not? And, how should such decision making occur?

Consider similar questions when applied to an employer-employee scenario. Does a third party's right to protect against harm outweigh the individual's right to not know? Would an employee with inherited susceptibility for diseases resulting from chemical exposures have the right to work in the high-risk job because they want to? Would it make a difference if the employee relieves the employer of financial liability for any illness incurred directly from their work? Should the decision change if the employee is a genius whose work may truly benefit humankind, and he or she agrees to risk premature death for the opportunity to create monumental benefit?

Ethically acceptable decision making involves balancing the rights and obligations of the various stakeholders. Situational characteristics and ethical considerations will inform the weighting of various factors and ultimately the defensibility of whatever balance is determined to be fair. To illustrate this point, consider that there are inconsistencies not only in which newborn screens are mandated by which states, but also the manner in which screening results are disclosed. Differences in disclosure requirements can lead to drastically different results with the implication of gross injustice as typified in the following case.

Creason v. State Dept. Health Services, 957 P.2d 1323 (Cal. 1998) involved the fact that Creason suffered irreversible harm because newborn results were reported as normal. Creason's abnormally low levels of both thyroxine and thyrotropin, findings consistent with the presence of congenital hypothyroidism, were not initially reported to her pediatrician, because the state only mandated the immediate reporting of abnormal results in which thyroxine levels were low and thyrotropin levels were high. By the time the actual test values were received, the infant had sustained irreversible disability from failure to diagnose and treat. Had the actual results been reported initially, treatment would have started immediately and enabled Creason to lead a normal life. The family sued the state for negligence in reporting,[28] but lost on the grounds that no laws were broken even though preventable injury did occur. Apart from the many ethical issues arising from wide inconsistencies among state newborn screening programs, reporting requirements and the handling of important test results are keys to ethical conditions of disclosure. The cases above illustrate some of the many ethical concerns surrounding the control of personal genetic information, appropriate disclosure, and inconsistencies among state screening programs.

Public opinion polls indicate that people do want to know what their genome is, and although they report strongly favoring restrictions to keep genetic information strictly private and inaccessible to any person or institution except by prior consent, the public is not necessarily knowledgeable about the extent to which their privacy might be compromised or the extent to which they could exercise a greater right to control. In 1992, the March of Dimes poll reported that 57 percent of the public think someone other than the patient should know their genetic test result. Of these respondents, 98 percent felt that a spouse or potential spouse should know.[29] The Congressional Office of Technology Assessment 1990 survey of employers found that only 12 of 330 fortune 500 companies were conducting genetic screening, though companies reported that they would be more likely to use testing once the cost decreased. Some 40 percent of the companies admitted that an employee's health insurance costs could affect their chance of employment.[30] A decade later, the public still wants to know their genetic profile, but it appears to be more fearful of adversity resulting from third-party disclosures. Two 2000 polls[31] found that the overwhelming majority of those asked wanted to have a genetic test and know their genetic risks. People reported that they were interested in multiplex testing for serious disease, even if there was no available treatment and were willing to pay out of pocket (up to $314). When asked their opinions about genetic privacy, 90 percent said that they'd want their test result shared with their doctor, and 69 percent would share their result with an unknown doctor to prevent serious disease. Regarding insurance, 17 percent believed that employers who offer health insurance should have access, whereas only 25 percent believed that life insurers should have access and 39 percent believed that health insurers should have access.

A 2002 Harris poll revealed that when given a definition of a genetic test, 81 percent of those asked reported that they would find taking a genetic test beneficial.[32] The same poll indicated that people report that they do understand what genetic testing is, and perhaps more troubling, recent research indicates people still do not understand what it is. Results showed that the public's level of knowledge about genetics has actually dropped in the past decade, despite the plethora of media reporting and genetics education efforts.[33] Arguably, the greatest challenge to protecting genetic privacy is the public's understanding of genetic information.

Genetic Discrimination Today

A recent study indicates that people are genuinely concerned about genetic discrimination.[34] Nonetheless, stakeholders have hotly debated whether any genetic privacy violations or genetic discrimination have in fact occurred as well as the likelihood of future instances. Some maintain that only unsubstantiated anecdotal reports exist and, therefore, much of the worry is misplaced. Others argue that bona fide discrimination has in fact occurred in a wide variety of contexts and that the threat of future abuses is real and must be contained. Many acknowledge that obtaining objective documented evidence of abuses is inherently problematic because of self-selection bias. Individuals are unlikely to come forward to report events, particularly if they believe that doing so may result in even greater adversity. These and other compelling factors that understandably deter victims from challenging occurrences make it nearly impossible to get an accurate read on incidence, much less, legal redress. Furthermore, the fact that discrimination is often covert, makes independent verification nearly impossible. Moreover, attempts to objectively measure incidence rely on research design assumptions that limit the accuracy of the research results. For instance, researchers have hypothesized that if genetic discrimination truly exists, it would have been reported to governmental authorities, such as State Insurance Commissioners. This hypothesis presumes (falsely) that injured individuals would seek redress through this mechanism. The study's finding that genetic discrimination did not exist, because the Commissioners lacked records of any cases, is refutable in light of evidence indicating that individuals did not seek recourse through their State Insurance Commissioner. People did not realize that the Commissioner handled grievances pertaining to insurance issues beyond automobile coverage, or they believed that because the Commissioner is obligated to investigate a claim, the insurers would learn of a forthcoming complaint and bolster their case against the consumer. Another research design complication consists in the fact that identifying actual cases requires self-reporting (the purported victims recounting their experiences). Many have argued that any study will be limited by self-selection bias. An objective randomized study is simply not feasible.

A belief in the real possibility for genetic discrimination has out-weighed the need to document its existence, because the majority of states have enacted genetic privacy protection and genetic discrimination prohibition legislation. While nearly every state in the United States has anti–genetic discrimination laws in place, it is noteworthy that the definitions and extent of enforcement terms differ considerably.[35] On the federal level, limited uniform protection exists based on the 1996 Kennedy-Kassenbaum law prohibiting genetic discrimination of individuals with preexisting disease within group health insurance plans. The federal workforce is protected against genetic discrimination under President Clinton's 2000 executive order. While no other broadly encompassing federal legislation outlawing genetic discrimination exists as of this writing, except for possible protection under the ADA, numerous bills have been and continue to be introduced, with some passing in the House or Senate, only to be shot down. As of this writing, GINA (the Genetic Information Nondiscrimination Act [H.R. 493, S. 358]) is pending. This federal legislation is designed to protect privacy and prohibit discrimination with the goal of promoting consumer uptake of DNA-based testing and access to related clinical benefits. The bill, originally proposed 12 years ago (by Representative Louise Slaughter, New York), bars health insurers and employers from refusals based on predictive genetic information, and from requiring genetic tests as a basis for consideration. It is expected to pass.

Existing legislation demonstrated a primary concern about abuse by insurers and employers. Concern about authorized use of genetic information extends also, though, to DNA banked samples, which theoretically could be used for purposes other than those for which the samples were originally obtained and consented to, as the samples can remain testable indefinitely. National advisory committees have addressed this concern.[36]

Personal Accounts of Genetic Discrimination

To illustrate the meaning of genetic discrimination, it is useful to examine individuals' reported experiences. The first federally funded national study of genetic discrimination was conducted in 1992–1994.[37] It sought to identify and document the breadth of situations in which individuals had experienced exclusions based on known or presumed genotype. This study was designed with the following aims: (1) to identify a range of social institutions that engaged in discriminatory practices, including but not limited to insurance companies and employers; (2) to evaluate the impact of a discriminatory experience on individuals and their families; and (3) to determine the underlying basis of the discrimination, that is, ignorance or institutional policy. The distinctive feature of this study is that it sought to distinguish genotypic- from phenotypic-based exclusions. As such, it investigated the possible genotypes within selected single-gene disorders that could provide a basis for carrier status (individuals who themselves are normal and will not develop disease), at-risk status for a late onset disorder (based on known family history), and successfully treated genetic conditions (individuals who as a result of treatment are asymptomatic and are at no greater risk for morbidity or mortality than the general population). This study's findings merit attention because this is the only study that identified a broad range of discriminatory contexts based on genotype, not phenotype.

Researchers found that genetic discrimination reportedly existed in settings so diverse and involving issues integral to personal life choices and quality of life decisions that it suggested anyone could experience genetic discrimination, in any facet, and at any point (each of us carry 8 to 25 mutated genes, which if present in two copies are capable of expressing deleterious effects). As noted in chapter 1, Dr. Francis Collins, current head of the NIH's National Human Genome Research Institute, and others frequently address this point when urging for prohibitive legislation and public policy. Findings include reports of individuals who are obligate carriers, and at risk for late onset disease, being excluded from access to many types of insurance (health, life, disability, long-term care, and even mortgage insurance) and refused coverage for preventative and curative treatments based on "preexisting condition" exclusions. In regards to the treated genetic condition, insurers were reported to refuse to cover these preventative treatments, despite their comparatively minimal cost and, employers were reported as basing adverse personnel decisions on the presence of a genetic diagnosis. The broad scope of contexts in which discrimination reportedly prevailed included the military, blood donation, adoption, education, reproductive choice, and one's family.

Note that this study[38] was conducted before **HIPAA** regulations, Kennedy-Kassenbaum, and the ADA enforcement guidance all went into effect. In addition, the majority of reported cases occurred before the institution of the ADA. Most of these insurance cases involve small-group coverage and self-insured groups, which are exempt from prohibitory laws. They are, however, illustrative of the real-life impact of such exclusions and it is worthwhile to review several of the study's cases.

Hereditary Hemochromatosis

One of the most frequent disorders excluded from insurance coverage was hereditary hemochromatosis. Hereditary hemochromatosis (HC) is an autosomal recessive iron overload condition that affects roughly 3 of every 1,000 Caucasian Americans, and currently represents the most common genetic disease in the United States. It is characterized by a long asymptomatic period before presenting clinically. Individuals typically do not manifest symptoms until they are well into their 40s. If undetected, clinical symptoms lead to serious and irreversible secondary disease, including cirrhosis and cancer of the liver, diabetes, cardiac disease, and arthritis. Though difficult to diagnose because early symptomatology is vague and masks as other conditions, it is easy and comparatively inexpensive to treat. Successfully treated individuals are those who have been treated before end organ failure has begun caused by iron overload. As such, they are asymptomatic and free of progressive secondary disease. Successfully treated individuals have a normal actuarial life expectancy. In other words, those treated for HC are at no greater morbidity or mortality risk than the general population, yet they reportedly were excluded from insurance coverage, employment opportunities, and the ability to donate blood because of their genetic diagnosis. Insurance refusals included health (individual, small group, large employment-based coverage, and medically necessary treatments), life, mortgage, supplemental, and disability.

Jane X., a young Midwestern woman in her late 20s, was approved for full standard health insurance as a dependent on her husband's small-group policy, while simultaneously undergoing a diagnostic workup. Four months later, when she was diagnosed with HC, her insurance company rescinded her policy on the basis of preexisting diagnosis. She was pregnant at the time. Tim Y., of South Carolina, was accepted for independent standard coverage, but dropped nine months later when he was diagnosed with the same condition, also on the basis of preexisting condition. Linda W., in Colorado, tried unsuccessfully to obtain health insurance for six years subsequent to her controlled HC diagnosis. Milton A., of Texas, was diagnosed while covered under a group policy. He dropped his policy to get more favorable coverage under his wife's policy. When she switched jobs, he tried to renew his former policy but was rejected because of HC. Older but nonretired individuals have similarly been refused long-term or supplementary health insurance. At age 63, 10 years after having been diagnosed with HC and in excellent health having successfully treated it, Mary C. was refused nursing home insurance because of her controlled HC diagnosis. Robert T. was refused group disability coverage because of his "blood condition." The rejection letter stated that the company does not take substandard risks, even though it is a group policy; however, his condition was successfully treated, meaning that he was not at greater risk of morbidity or mortality. Jim O., at the age of 58 and diagnosed with HC, wanted to start his own business. When inquiring about the feasibility of obtaining health insurance independently, several carriers refused him on grounds of his HC. He gave up his dream of owning a business. He wanted to take advantage of an early retirement program, but fearful of insurance exclusion, opted to continue working. In all of these cases, despite letters from people's physician attesting to their patients' good health status, excellent prognosis, and the fact that HC did not constitute a health risk, insurers refused to reverse their exclusionary decisions.

Bill P., in Pennsylvania, was able to keep his insurance, but the insurer refused to pay for phlebotomy treatments unless it was provided on a hospital inpatient basis, which incurred significant unneeded hospital charges. If Bill P. was able to have his treatments covered by going to a local blood bank, the costs involved would be greatly lower. However, the local blood bank refused his blood because of his diagnosis. Furthermore, the Red Cross and affiliated blood banks had, until very recently, refused the blood not because of safety risks or labeling requirements but because donations must be altruistic. In as much individuals with HC give blood out of medical necessity, blood banks refused donations because they were based on self-interest, not altruism, and did so despite occasions of national blood shortages. While donated blood must be labeled according to federal law and regulations, physicians had been reluctant to use HC blood for transfusions, fearing patient upset—or worse, malpractice suits based on safety concerns related to receiving the blood of an individual with a genetic disease. Within the past couple of years, the Red Cross prohibition was reversed to permit the use of HC blood.

Porphyria

Another example of the type of exclusions that are likely to become more common because of their phenotypic variability involves the condition porphyria. Porphyria is a group of autosomal-dominant inborn errors of metabolism resulting from multiple different enzyme defects, which can result in the production of abnormal metabolites. Because environmental factors play a major role in the clinical expression of all forms of the disease, many gene carriers stay asymptomatic for years or never become symptomatic. Though particular types are more prevalent in specific populations, the incidence in the United States is estimated to be 5 to 10 per 10,000. Kathy L. has a family history of porphyria but, no clinical evidence that she herself has the condition. Once she did experience a mild episode, though this event was classified as an incidental finding, was unrelated, and no treatment was indicated. When applying for employment-based health insurance, the insurer learned that she had been tested for the condition and this raised suspicion. Although her results were inconclusive, not ruling out but not confirming the diagnosis, Kathy was refused insurance. Her physician assisted her in challenging the exclusion, arguing that Kathy did not have porphyria. She was offered insurance only after complying with a 12-month waiting period for preexisting condition, just as if she had a verified diagnosis.

Huntington Disease

Huntington disease (HD) is an autosomal dominant and progressive neurodegenerative disorder with early mortality and an inheritable risk to offspring of an affected parent being 50 percent. Wayne C., in Colorado, was repeatedly refused health insurance based on his family history of HD, even though he may never become ill. Paula S. was denied private supplemental long-term care insurance unless she agreed to undergo genetic testing to determine whether she had the gene. The insurer refused to pay for testing, and Paula did not want to be tested because she believed (falsely) that the test was unreliable and inconclusive. Moreover, she did not want to know her results, regardless of the test's reliability. Patients with HD have been excluded from renal transplantation lists on the basis that HD made them undesirable recipients.

Phenylketonuria

Phenylketonuria (PKU) is an autosomal recessive disorder resulting from a deficient enzyme. It is, however, a treatable condition and newborn screening is widely mandated because dietary treatment prevents the accumulation of the toxic metabolite and subsequent mental retardation. Yet, dietary treatments are frequently not covered by insurers. Moreover, many infants with PKU have been refused health insurance because of their preexisting condition.

Exclusions of Unaffected Blood Relatives

In many cases, discrimination extends beyond the individual to their relatives, that is, family members *without* genetic diagnoses or risks. For example, Jerry M., of Kentucky, worked for a small company of five employees. When the company renegotiated its benefit package, she was informed that she was ineligible for insurance because of her husband's HC diagnosis, even though he was successfully treated and had been asymptomatic for several years. Similarly, his large and self-insured employer informed Ron S. that he was ineligible for health insurance because of his wife's HC diagnosis. In Massachusetts, an entire family of four was refused health insurance because one of the two children in the family had PKU, consistent with the parents being unaffected carriers.

Some individuals at risk of HD reported that life insurers required genetic test results as a condition of eligibility. While some people may seek genetic testing to obtain negative results to prevent exclusion, those who decline testing have reportedly been perceived as too great an insurance risk. Furthermore, even negative genetic tests results have been shown to be insufficient to prevent exclusion, presumably because the mere act of undergoing genetic testing is perceived as likely to indicate a serious problem. Here, a negative result is deemed to be uncertain and unreliable, hence irrelevant, information.

In the area of employment, individuals reported a failure to be hired or attain promotion, job loss, and even job lock on the basis of their genotype. In one case, a man with a family history of HD reported failure to be promoted from midlife on. A man with controlled HC reported the failure to obtain a job as an insurance salesman, because a manager considered his diagnosis too great a risk, despite the fact that it had no bearing on his ability to perform job responsibilities. In another case, a woman was summarily fired after she inadvertently disclosed her familial risk for HD. Many employment issues were framed in the context of employer health benefit concerns.

In the area of adoption, individuals at risk for HD and carriers of muccopolyssacharidosis reported that agencies required genetic testing as a condition of application consideration. Muccopolyssacharidoses syndromes constitute a heterogeneous group of lysosomal storage diseases in which mucopolysaccharides (or glycosaminoglycans) accumulate in lysosomes as a result of a deficiency of one of several lysosomal enzymes required for their degradation. Two of these conditions, Hunter syndrome and Hurler syndrome, first described in 1917 and

1919, respectively, were originally called "gargoylism," because the facial features of affected individuals appeared coarse. Affected children are mentally retarded, and have skeletal abnormalities, short stature, and a host of other abnormalities. No effective treatment currently exists and affected individuals have a typical life span of 11 to 13 years.

In one anonymous case, a birth mother in the early stages of HD was refused the opportunity to put her infant (at risk for HD) up for adoption with a private agency but was accepted by a public agency. A couple, one of whom was at risk for HD, attempting to adopt had been similarly refused by a private agency, but later was accepted by the same public agency to adopt this at-risk baby. Whether one agrees with the decision, the agency determined that these parents were unfit to parent a normal child, but acceptable to parent an otherwise presumably disabled child who possibly, but not certainly, was presymptomatic. The agency's decisions suggest, furthermore, that children at risk for disabilities—even if late onset—do not have the right to as "good" a parent as those children who are not at risk.

In the area of education, healthy siblings of a child diagnosed with Hurler syndrome (one of the mucopolysaccharidosis disorders; a lethal condition resulting from a severe deficiency of the alpha-L iduronidase enzyme) reported mistreatment and inappropriate detention by school teachers and a principal, who incorrectly believed that they were likely to become ill. Alternatively, children with mucopolysaccharidosis without cerebral disease, in particular Morquoio syndrome, have been inappropriately placed in special learning programs under the misapprehension that they were mentally retarded. In other cases, individuals at risk for HD reported failure to obtain professional training (a medical residency and graduate school) because of their genetic risk.

In the area of health care delivery, pregnant women, with an autosomal dominant condition in their family, reported pressure from clinical geneticists, obstetrician/gynecologist physicians, and pediatricians to undergo prenatal genetic testing, abort an affected fetus, and avoid future reproduction. In one case, a woman whose husband had HD reported being coerced into genetic counseling and testing as a condition of maintaining prenatal care. A pregnant woman whose only child had PKU was strongly advised by her pediatrician to not reproduce and abort the pregnancy. In another case, a pregnant mother was threatened by her health insurer with refusal to pay for the delivery and care of her baby if he/she was diagnosed with CF. The insurer tried to coerce her into undergoing prenatal testing, though she refused. At the time, the test identified only a few hundred of the then-known 300 mutations.

We now know that there are well over 900 CF mutations, though the test only examines 251 of the most common ones, making the possibility of obtaining a false-negative result quite real. In this case, false assurance of the absence might have satisfied the insurer, at least temporarily.

These cases illustrate some of the many ethical concerns arising in the provision of health care, such as possible coercion to undergo genetic testing, conflicting rights and obligations, the ethical defensibility of trying to control against the birth of unhealthy individuals, and the appropriate use of new technologies given their limitations. An ongoing concern involves ethical issues arising from test results that in the future may mean more than what was originally consensually tested. For example, some genes are both deleterious and protective, as in the case of the sickle cell gene, which protects against malaria in environments where it is prevalent.

Both men and women at risk for HD reported failure to be admitted into the military because of their risk status, despite the exceedingly low probability of becoming ill during service. Some individuals reported being admitted to the military with full knowledge of their risk, but were later discharged without disability benefits on grounds they were disabled, again based on their HD risk.

Circumventing Genetic Diagnosis

Of particular interest are instances in which individuals acted to avoid the possibility of genetic discrimination by preempting a genetic diagnosis. Foreseeing possible exclusion, some individuals asked insurance agents to do preapplication screens of their insurability. Others, presuming themselves to be presymptomatic, restricted their life opportunities on the basis of their assumption of inevitable disease.

The extent to which individuals may be motivated to avoid genotypic-based discrimination is illustrated by the following case. The two children of a man diagnosed with treated HC self-prescribed phlebotomy regimens for themselves with the same regularity as their father's treatments to prevent a genetic diagnosis and thereby circumvent exclusions. This case is particularly interesting because these children, as the offspring of only one parent with the condition, were at only negligible risk of becoming ill; however, they acted as if they harbored an undetected diagnosis despite the fact that HC's mode of inheritance made it highly improbable that they would be affected.

In many cases, once a genetic diagnosis is established, patients seek to share this information with blood relatives, even those disconnected from the family. In some but not all instances, news of genetic diagnosis is far from welcome and evidence exists of intrafamilial ostracization.[39]

The Limitations of Assurance of Privacy

In addition to reported instances of genetic discrimination, an array of reported privacy invasions exists. U.S. law requires that individuals manage the privacy of their genetic information. Many of these reported concerns resulted from voluntary self-disclosures and not third-party access, so the concerns hinged on questions of presumed confidentiality. Some individuals claiming to have been discriminated reported lacking the awareness that voluntary self-disclosures (sharing their genetic status with coworker friends) could adversely affect them. For example, private information disclosed in the context of a friendship, and presumed to remain confidential, was passed on to insurers and employers who used that information to the individual's detriment. In several instances, people sought insurance policies from their friends who were agents, but did so unaware of the fact their friends were professionally obligated to use information obtained on a friend-to-friend basis if it had direct bearing on the insurance company's underwriting interests. In another case, a couple was accused of dishonesty by an adoption agency because they preferred to disclose the wife's HD risk only in a personal conversation with agency personnel and not on the biographical statement required as part of the initial application process. Also, potentially contentious are family history disclosures. Knowing the importance of early risk identification, physicians are increasingly obtaining family history information from their patients. Patients disclose that family history, even though for the most part their family members have not consented to such disclosures. However, requiring documented familial consent before disclosure during routine care is not the norm and, as a practical matter, would be overly burdensome. In summary, voluntary but unintentional disclosures can occur when people are unaware that communicating personal or familial genetic information could implicate themselves for future denials of opportunities or societal benefits.

Involuntary disclosures resulting in exclusions can occur as well. These types of disclosures can result from access to public records or common knowledge in small rural towns. In one case, a man at risk for HD reported failure to obtain what would have been otherwise routine promotions after a supervisor inferred his risk status from reading a relative's obituary in the newspaper. These types of disclosures challenge our ability to safeguard the privacy of one's genetic information.

Privacy Protection Has Its Limitations

Learning one's genetic information can result in positive or adverse effects. People have reported that knowing their genetic information has had a profound effect on their life choices. Some have reported lowered self-esteem and perceived restrictions on career development and reproductive freedom (some felt unworthy of marriage and reproduction). Notably, the impact can extend to family members as well. Intrafamilial discord has ensued when family members blamed each other for problems resulting from a genetic diagnosis.

The extent to which such events adversely affect individuals is highly idiosyncratic, given the vast array of factors that influence an individual's psychosocial experience. While policy efforts have focused on protecting against genetic discrimination primarily in the context of insurance and employment, the potential for genetic discrimination extends far beyond these venues. Moreover, questions about parental or state ownership of a child's genotype also raise privacy concerns. Case law suggests that the government has a right to mandate genetic testing of children; newborn screening is a case in point. The speed with which tests are becoming commercially available raises questions about the potential government involvement in other types of genetic testing, including prenatal testing, diagnostic testing of children, or predispositional screening of large adult populations. Consider a relatively recent case involving a state's interest in using genetic testing as a mechanism to reduce fiscal burden. In 1993, researchers in Colorado and Georgia conducted genetic tests on children in special education classes in selected public schools.[40] The genetic screening programs were designed to identify children affected by Fragile X syndrome, which is an inherited form of mental retardation. In Colorado, researchers justified the screening program on economic grounds, citing reports that Colorado spends $1,609,852.63 over the lifetime of a person with Fragile X syndrome, over and above expenditures for so-called normal children. They extrapolated these estimated costs to a $4 billion burden for Colorado in caring for these affected children, and a national burden of $280 billion. Researchers argued that large-scale screening coupled with reproductive counseling could relieve this burden by one half. Government interference with individuals' rights is generally discouraged because it is inconsistent with present norms about its role, but shameful decisions in legal history, such as forced sterilization laws, evince otherwise. In as much as the fiscal burden of Fragile X was felt sufficient to consider mandated screening, the potent example of the state's right to enforce coerced testing is illustrated by conscripted syphilis testing before obtaining a marriage license in the late 1930s early 1940s.[41]

Ethical principles and legal doctrines upholding respect for the autonomy of an individual are well established in the United States, though not without occasions of conflict. At issue in decisions about generating genetic information, and then safeguarding it down the road, is the meaning and value of the *autonomy* of the individual decision maker. In regards to children, and other people deemed incapable of making their own decisions, there are concerns about whether the applicable laws or surrogate decision makers really do protect the individual's interests, especially in light of the fact that genetic testing decisions are likely to have profound implications for those who are tested.

An expansion of children's rights—which began in the early 1970s, and arguably arose from social circumstances deemed likely to result in adversity to a child and from the fact that alternative mechanisms to protect children had failed—may be threatened by the expanding availability of new genetic tests. The adequacy of existing law to provide a satisfactory forum for structuring moral and legal conflicts regarding whether children ought to undergo genetic testing (separate and apart from the standard of care that they should not be tested unless they can directly receive a medical benefit from doing so) has not yet been tested. Furthermore, it may be inadequate to protect the interests of children from decisions made by parents or the government. Affording parents an interim ownership of their child's genetic information (until the child reaches adulthood, as designated by state law) in the name of the child's best interest could provide a basis for a legal dispute for which children may have no recourse.

Viewing these cases in the context of public opinion polls is interesting because it not only illustrates potential discrepancies between what people believe and what they actually do, but also the practical difficulties in protecting privacy and proactively guarding against abuse. Two recent polls (Dr. Koop and *Harris*) found that the majority of those asked overwhelming wanted to have a genetic test and know their genetic risks. People reported that they were interested in multiplex testing for serious disease, even if there was no available treatment, and were willing to pay out of pocket up to $314 (in 2002 dollars). When asked their opinions about genetic privacy, the response was that 90 percent wanted their test result shared with their doctor; 69 percent would share their result with an unknown doctor to prevent serious disease; and 17 percent believed that employers who offer health insurance should have access, whereas 25 percent believed that life insurers and 39 percent believed that health insurers should have access.[42]

Fairness and Use of Genetic Information

In the United States, it is well established that medical information is strictly confidential and state laws restrict access to medical records. Many view this limitation as insufficient to guard against unauthorized access and unfair use of genetic information. But what is fair use of genetic information? Who ought and ought not to have access to personal genetic information and why? What uses ought to be sanctioned legally and morally, and which not? Difficulty in resolving these questions lays in part in blurry distinctions between medical and genetic information. For example, a cholesterol test result indicates a cholesterol level that functionally provides a measured risk of cardiac disease. An Apo enzyme E gene test result similarly provides an indication of a level of risk for cardiac disease. Given the similarity in these two tests, some argue that it is unfair to permit insurer access to the former while barring access to the latter. Insurers, in particular, have argued that genetic tests are really nothing more than medical tests that insurers already have access to so that the may underwrite insurance policies. Furthermore, they claim that if they are prohibited from using genetic information (which has bearing on establishing an individual's actuarial risk), they are likely to be cheated by people who will load up on insurance knowing they are at high risk. Insurers maintain that this will cause exorbitant hikes in premiums for all, if not force insurers out of business. Public interest groups, by contrast, fear that insurers will exclude anyone whose genetic test indicates anything short of a genetically clean bill of health. They claim that the adverse societal consequences of creating a genetic underclass could be significant, because the effects of exclusion go well beyond insurability.

In the greater scheme of things, it is unlikely that genetic information will *not* matter. The important questions are how much ought it to matter? How much weight ought we to give genetic information? Should genotypes matter more than phenotypes? Is differential treatment based on genetic information ever morally justified, and if so, then on what grounds? Should utilitarian concerns, such as the greatest good for the greatest number, provide a rule of thumb? If so, then what is the definition of the greatest good, and who makes that determination? Should principles of distributive justice—that is, the notion that social benefits and harms are fairly distributed by treating like cases the same way and unlike cases differently—determine how significant genetics should be and in which contexts? Or, can a more compelling case be made for principles of care to guide ethical decision making?

By what ethical principles, or guidance, do we as a society assign weights to the potential burdens and benefits of using genetic information? How do we weigh such burdens against such benefits when the scale involves weighing quantities versus qualities? For example, is it fair for society to expend scarce resources on individuals who are likely to give back only little, if anything, to society? Should social policy limit the extent to which people can benefit from their genetics to maintain equal opportunities? To what extent (if any) ought economic considerations factor into social policy? Should genetic risk provide a basis on which to limit individual rights to protect the public from harm? For example, individuals with Long Q T syndrome (sudden unpredicted death caused by heart failure) are unlikely to be permitted to pilot commercial airplanes, though such individuals are rarely identified before their sudden death. Yet, the truth of the matter is that a genetic risk is only risk and not an inevitability. The tendency to reify genetics creates the tendency to give undue weight to such factors.

Answers to these questions require deciding what we mean when we use genetic status as the basis for differential treatment. But even this application is far from clear. Are we referring to the specific genotype, as revealed by testing, which may or may not have a clinically visible expression? Or are we referring to genotype and its corresponding phenotype? Medicine's focus has been blurry. For example, in offering prenatal diagnosis for certain diseases, society appears to support genotype and phenotype prevention. In this regard, we seem to discriminate against those who chose to bring a child with CF into the world, by sanctioning the termination of an affected pregnancy as rational or even desirable. In permitting termination because of CF, we are in effect saying that a CF genotype ought not be perpetuated and a CF phenotype ought not exist because of its burden to the individual affected and his/her parents, if not also to society. Research efforts likewise appear to discriminate against affected individuals by focusing on genotype prevention and not on disease prevention or disease management. Unlike single-gene disorders, the vast majority of testable conditions will be multifactorial, conditions resulting from gene-gene and gene-environment interactions. Testing for multifactorial conditions (e.g., hereditable risk for cardiovascular disease, diabetes, or types of cancers) is aimed at phenotype prevention, that is, preventing disease symptomatology. Taken together, society's approach can be viewed as inconsistent: endorsing genotype prevention on the one hand and phenotype prevention on the other.

In truth, we are at the precipice at which we need to determine consistent, coherent and most important ethically justifiable policy and practices. If the overriding goal of genetic testing and screening is to give us the best medicine possible, then arguably we are oriented toward the philosophy that phenotype prevention is appropriate and would be consistent with the goals of medicine, namely, to produce the best preventive personalized medicine possible.

Furthermore, when gene expression is not obvious (such as a low hematocrit) will such things as a biochemical level be used as a basis for differential treatment? Moreover, are all genetic traits morally relevant to issues of fair treatment? What weight should be given to genetic traits versus other morally relevant factors, such as the inherent dignity of all human beings and the moral requirement to treat people with a fundamental respect for their humanity? Should genetic factors be given greater weight than other biological factors or other nongenetic factors? Ought *fairness* be determined by particulars unique to each context, so that genetic differences in height might be morally relevant in some situations but not others? For argument, say that it is fair to charge a short person higher automobile insurance rates than a tall person, if short people cause statistically more automobile accidents that include fatalities. Any short-stature driver would be deemed statistically a greater risk on the road than any given tall driver. Statistical averages might indicate what may be true of most short drivers, but they would not necessarily prove that any particular driver, who is short, is a significant liability on the road.

Other questions exist about access to genetic information. In practice, there is concern given the escalating cost of health care and the increasing numbers of the uninsured that only the wealthy have access to genetic tests, and the potential benefits resulting from the results, because they can afford to pay for them and for treatments that are not otherwise covered by insurers or state programs. Clearly, such a scenario is far from just and arguably is ethically indefensible.

Access remains a potent concern, not only in the United States but globally. Whether people in developing countries will have access to benefits available to those in developed countries is a significant ethical concerns, particularly in light of scarce resources and health conditions that are already ethically suspect. Future innovations may create additional concerns. Will the much-fantasized genotype at birth cards become a reality, based on a legal requirement for every newborn, thereby rendering state ownership of a newborn's DNA? Ought genotypes be identified for children, even though they are not old enough to legally give consent and may not want to know their results? Will advances that inform reproductive decision making ultimately lead to the obsolence of certain genotypes? Professional organizations and policy makers have recommended that testing be permitted only if there is clear and direct actionable medical benefit, yet the benefit to whom (whose rights predominate, the parent's or child's?) is ethically debatable.

As a theoretic matter, a key question is whether moral theories of duty compel us to know the future or even provide a justification for why we should want to know our genetic profile.[43] As a practical matter, these ethical questions are important not only for purely moral or legal considerations, but also for social and public policy. It is important to note that other factors, such as economic and political interests and not ethical considerations alone, will shape policy. Ensuring the just treatment of individuals will involve deciding the ethically appropriate weight to be given to genetic factors, lest we reify genetics as the sole arbiter of an ethical decision making.

To be ethically defensible, policy decisions will need to cohere with concepts laid out in the Declaration of Independence, which decree what one is entitled to in virtue of the fact that they've been born. In stating that all men are created free and equal, we have determined that biological features (i.e., color of an individual's skin, gender, etc.) do not undermine inherent equality. How societies frame the questions that will weigh competing values ultimately will determine policy. Moreover, social policies, while representing societal value choices, do not necessarily result from an informed citizenry or democratic processes, which is another factor to weigh in the ethical debate about appropriate policy formation. Whether choices will enhance or diminish the value attached to personhood remains uncertain and may well be determined more by markets than morals.

Notes

1. Raithatha, N., and Smith, R., "Disclosure of Genetic Tests for Health Insurance: Is it Ethical or Not?" *Lancet* 363 (January 31, 2004).

2. President Clinton executive order bans genetic discrimination, http://usgovinfo.about.com/library/news/aa020800a.htm; also see http://eeoc.gov/press/4-18-01.html.

3. *Burlington Northern Santa Fe Railway Co. v. White.* 364 F.3d 789 (6th Cir. 2004); Lower Court Case Number: 00-6780, 01-5024; see www.supreme courtus.gov/qp/05-00259qp.pdf.

4. Genetic discrimination in this context is defined as exclusion based on genetic information (actual genotype, perceived genotype, or negative effects arising from a genotype); see www.cdc.gov/genomics/gtesting/ACCE/FBR/CF/CFGlossary2.htm#Glossary.

5. Adoption was out of the question, because agencies typically exclude individuals who are at risk for Huntington disease. The rationale underlying such refusals is that the risk of severe parental disability and even premature death is too great for any child. Notably, the child's age at the likely time of onset of the parental disease is not a factor. The argument extends further, that orphaned children ought not to be given to adults whose functionality or life span is likely to be shortened by severe disability or death, particularly when there are plenty of potential couples who are fit to parent and have no such liabilities. While it could be possible for such an at-risk couple to rear a child who would be a fully independent adult by the time the affected parent develops disease, this possibility is not considered. Similarly, those with a family history, in the absence of a positive gene test, may not have inherited the gene and thus never develop disease, and yet this possibility is not considered. It is possible that an agency could give a child to a "normal" couple who, owing to the vicissitudes of life, could suddenly die from other causes (car accident, sudden heart attack, etc.), leaving the child again orphaned, but risk-based probabilities do not factor into agency decision-making policy.

6. Davenport, C. B., *Heredity in Relation to Eugenics* (New York: Holt, 1911).

7. Laughlin's Model Eugenical Sterilization Law was published in 1922 and served as a basis for lobbying Congress to restrict immigration. These efforts resulted in the Johnson Act of 1924, which imposed strict quotas on immigration. Jews, in particular, were allotted low quotas. Harry Hamilton Laughlin, D. Sc., Assistant Director of the Eugenics Record Office, Carnegie Institution of Washington, Cold Spring Harbor, Long Island, New York, and Eugenics Associate of the Psychopathic Laboratory of the Municipal Court of Chicago, Published by the Psychopathic Laboratory of the Municipal Court of Chicago, December 1922.

8. Bird, R. D., and Allen, G., "The J.H.B. Archive Report: The Papers of Harry Hamilton Laughlin, Eugenicist," *Journal of Historical Biology* 14, no. 2 (1981): 339–53.

9. Allen, G. E., "The Misuse of Biological Hierarchies: The American Eugenics Movement, 1900–1940," *History and Philosophy of the Life Sciences* 5 (1983): 105–28.

10. Bregin, P., and Bregin, G., *The War Against Children: How the Drugs, Programs, and Theories of the Psychiatric Establishment Are Threatening America's Children with a Medical "Cure" for Violence* (New York: St. Martin's Press, 1994).

11. *Wrestling with Behavioral Genetics: Ethics, Science, and Public Conversation*, ed. E. Parens, A. Chapman, and N. Press (Baltimore: Johns Hopkins University Press, 2005).

12. Neel, J. V., Wells, I. G., and Itano, H. A., "Familial Difference in the Proportion of Abnormal Hemoglobin Present in Sickle-Cell Trait," *Journal of Clinical Investigation* 30 (1951): 1120; Beet, E. A., "The Genetics of the Sickle Cell Trait in Bantu Tribe," *Annals of Eugenics* (London) 14 (1949): 279; Neel, J., and Shull, W., *Human Heredity* (Chicago: University of Chicago Press, 1954).

13. Konotey-Ahulu, F. I. D., "Effect of Environment on Sickle Cell Disease in West Africa: Epidemiologic and Clinical Considerations," in *Sickle Cell Disease, Diagnosis, Management, Education and Research*, ed. Abramson, H., Bertles, J. F., and Wethers, D. L. (St. Louis: C. V. Mosby, 1973), 20.

14. Gary, L., *Black Men* (Newbury Park, CA: Sage, 1981); Gary, L., "Social Research and the Black Community: Selected Issues and Priorities: A Selection of Papers from a Workshop on Developing Research Priorities for the Black Community," Institute for Urban Affairs and Research, June 25–29, 1974; Faden, R., Geller, G., Powers, M., eds., *AIDS, Women, and the Next Generation: Towards a Morally Acceptable Public Policy for HIV Testing of Pregnant Women and Newborns* (New York: Oxford University Press, 1991).

15. Beutler, E., et. al., "Hazards of Indiscriminate Screening of Sickling," *New England Journal of Medicine* 285, no. 26 (December 23, 1971): 1485.

16. Culliton, B. J., "Genetic Screening: States May Be Writing the Wrong Kind of Laws," *Science* 191, no. 4230 (March 5, 1976): 926–29; Culliton, B. J., "Genetic Screening: NAS Recommends Proceeding with Caution," *Science* 189, no. 4197 (July 11, 1975): 119–20.

17. Crawley, D., and Ecker, M., "Integrating Issues of Gender, Race and Ethnicity into Experimental Psychology and Other Social Sciences Methodology Courses," *Women's Quarterly* 1–2 (1990); Hampton, M., et al., "Sickle Cell 'Nondisease': A Potentially Serious Public Health Concern," *American Journal of the Diseases of Childhood* 128 (July 1974): 58–61; Hill, S., "Motherhood and the Obfuscation of Medical Knowledge: The Case of Sickle Cell Disease," *Gender & Society* 8, no. 1 (March 1994): 29–47.

18. Elfagm, A. A., Gul, S., and Bughaigis, Y., "First-Cousin Marriage and Sickle-Cell Anaemia," *Cellular and Molecular Biology, Including Cyto-Enzymology* 26, no. 2 (1980): 109–10; Brown, R. E., Hamilton, P. J., Kagwa, J., and Warley, M. A., "Is Marriage Counseling Feasible in Africa to Prevent Sickle-Cell Disease?" *Clinical Pediatrics* 8, no. 7 (July 1969): 421–24; Ohaeri, J. U., and Shokunbi, W. A., "Attitudes and Beliefs of Relatives of Patients with Sickle Cell Disease," *East African Medical Journal* 78, no. 4 (April 2001): 174–79; Farber, M. D., Koshy, M., and Kinney, T. R., "Cooperative Study of Sickle Cell Disease: Demographic and Socioeconomic Characteristics of Patients and Families with Sickle Cell Disease," *Journal of Chronic Diseases* 38, no. 6 (1985): 495–505; Durham, W. H., "Testing the Malaria Hypothesis in West Africa," in *Distribution and Evolution of Hemoglobin and Globin Loci* (New York: Elsevier Science Publishing, 1983); Durham, W. H., *Coevolution: Genes, Culture, and Human Diversity* (Stanford, CA: Stanford University Press, 1991).

19. "Newborn Screening for Sickle Cell Disease and other Hemoglobinopathies," 6, no. 9 (April 6–8), http://consensus.nih.gov/cons/061/061_intro.htm.

20. *Oxford American College Dictionary* (New York: Oxford University Press, 2002).

21. Warren, S., and Brandeis, L. D., "The Right to Privacy," *Harvard Law Review* 4 (1890): 193–97.

22. Beauchamp, T. L., and Childress, J. F., *Principles of Biomedical Ethics*, 5th ed. (New York: Oxford University Press, 2001).

23. Weston, A., et al., "HLA-DPB1 ✳E69 and Chronic Beryllium Disease," HuGEnet Fact Sheet, April 10, 2003, http://www.cdc.gov/genomics/hugenet/factsheets/FS_Beryllium.htm; Casteleyn, L., and Van Damme, K., "Acceptability of Genetic Susceptibility in Occupational Health: A Position Paper," *Archives of Public Health* 62 (2004): 15–22; Holtzman, N., "Ethical Aspects of Genetic Testing in the Workplace," *Community Genetics* 6 (2003): 136–38; Rossman, M. D., "Chronic Beryllium Disease: Diagnosis and Management," *Environmental Health Perspectives* 104, suppl. 6 (1996), http://ehp.niehs.nih.gov/docs/1996/Suppl-5/rossmanabs.html.

24. In the 1970s, the American-based Cyanamid company established a policy excluding all women of childbearing age from working at numerous high-paying jobs at the plant on the grounds that the prohibition was necessary to avoid possible birth defects in offspring caused by exposure to industrial chemicals. These protection policies continued well into the 1980s, with federal courts deciding fetal protection policies at several companies, including Olin, General Motors, B. F. Goodrich, and several hospitals. The best known of these cases was Johnson Controls, the largest U.S. auto battery manufacturer, which established a policy of excluding fertile women from jobs that involved exposures. Women of reproductive age who wanted to keep their assembly-line jobs had to supply medical proof of sterility. In 1989, the federal court of appeals, 7th Circuit in Chicago ruled that Johnson had the right to exclude women. One judge on the bench who dissented estimated that 15 to 20 million would be excluded from high-paying jobs because of the decision. A broad coalition of women pursued the decision and on March 20, 1991, the U.S. Supreme Court ruled that the company did not have the right to exclude women, basing the decision on the Civil Rights Act of 1964 (see Schmidt, William E., "Risk to Fetus Ruled as Barring Women from Jobs," *New York Times*, October 3, 1989, A16.; Greenhouse, Linda, "Court Backs Right of Women to Jobs With Health Risks," *New York Times*, March 21, 1991, 1, B12).

25. *Tarasoff v. Regents of University of California*, 17 Cal. 3d 425, 551 P.2d 334, 131 Cal. Rptr. 14 (Cal. 1976) On October 27, 1969, Prosenjit Poddar killed Tatiana Tarasoff. In this case, the court found that the psychiatrist should have warned Tatiana Tarasoff that his patient (Poddar) had disclosed his intent to murder her.

26. Heidi Pate and James Pate, *Petitioner vs. James B. Threlkel*; Geessler Clinic, P.A.; Shands Teaching Hospital & Clinics Inc., and Florida Board of Regents.

27. *Safer v. Pack*, 677 A.2d 1188 (N.J. App.), appeal denied, 683 A.2d 1163 (N.J. 1996).

28. *Creason v. State Dept. Health Services*, 957 P.2d 1323 (Cal. 1998).

29. "American's Knowledge and Attitudes towards Genetic Testing," Louis Harris and Associates, the March of Dimes, 1992.

30. The Congressional Office of Technology Assessment, "Genetic Monitoring and Screening in the Workplace," 1990.

31. Times/CNN poll and poll conducted by Dr. Koop.

32. The Harris Poll, No. 26, June 5, 2002, Harris Interactive (1,013 respondents).

33. Center for Genetics and Public Policy, Survey of the Public's Attitudes and Knowledge about Genetics, 2002; http://www.dnapolicy.org/research/reproductiveGenetics.jhtml.html.

34. Kass, Nancy, "Individuals with Genetic Conditions Twice As Likely to Report Denial of Health Insurance than Individuals with Other Chronic Illnesses"; http://www.jhsph.edu/publichealthnews/press_releases/2007/kass_genetic_testing.html; Johns Hopkins Center for Genetics and Public Policy survey, May 2007; http://www.nationalpartnership.org/site/DocServer/FacesofGeneticDiscrimination.pdf?docID=971; http://www.bioethicsinstitute.org; and "Concerns in Primary Care Population about Genetic Discrimination by Insurers," *Genetics in Medicine* 7, no. 5 (May–June 2005): 311–16.

35. National Human Genome Research Institute, "Welcome to the NHGRI Policy and Legislation Database," http://www.genome.gov/Policy-Ethics/LegDatabase/pubsearch.cfm.

36. Steinberg, K. K., "Ethical Challenges at the Beginning of the Millennium," *Statistics in Medicine* 20, nos. 9–10 (2001): 1415–19, http://www.cdc.gov/genomics/info/reports/policy/ethical.htm (accessed March 2005).

37. Barash, C., "Genetic Discrimination and Screening for Hemochromatosis: Then and Now," *Genetic Testing* 4, no. 2 (2000): 213–18; Geller, L., et al., "Individual, Family, and Societal Dimensions of Genetic Discrimination: A Case Study Analysis," *Science and Engineering Ethics* 2, no. 1 (January 1996): 71–88.

38. Billings, P., Kohn, M., de Cuevas, M., Beckwith, J., Alper, J., and Natowicz, M. R., "Discrimination as a Consequence of Genetic Testing," *American Journal of Human Genetics* 50 (1992): 476–82.

39. Apse, K., et al., "Perceptions of Genetic Discrimination Among At-Risk Relatives of Colo-Rectal Cancer Patients," *Genetics in Medicine* 6, no. 6 (November–December 2004): 510–16; Van Riper, M., "Geneic Testing and the Family," *Journal of Midwifery and Women's Health* 50, no. 3 (May–June 2005): 227–33.

40. Billings, P., and Hubbard, R., "Fragile X Testing: Who Benefits? School-Sponsored Genetic Screening Program Raises Specter of Discrimination," *Gene Watch* 9, nos. 3–4 (January 1994): 1–3.

41. Note that as of 1996, however, 15 states still require premarital syphilis testing as a requirement for marriage licenses (CDC, Division of STD Prevention, unpublished data, 1996).

42. http://www.harrisinteractive.com/harris_poll/index.asp?PID=304; http://www.pandab.org/HealthSrvyRpt.pdf.

43. "Know the Future: A Kantian Perspective on Predictive Genetic Testing," *Medical Health Care Philosophy* 8, no. 1 (2005): 29–37. One can also ask the same question in the context of other moral theories summarized in Chapter 2.

CHAPTER 4

Mean Genes: Hereditability, Human Behavior, and a Genetic Basis for Aggression?

Why do economically successful parents tend to produce economically successful kids? Could it be that economic success is rooted in genes? Although less frequently discussed than disease-related genetic advances, no less significant are the advances in behavioral genetics. Recent findings indicate that genes influence a wide range of human behaviors and traits, such as (but not limited to) athleticism, taste, social phobia, intelligence, aggression, and recently even economic success.[1]

Researchers recently found that inequality in America is on the rise, meaning that there is less intergenerational mobility, suggesting that Americans' economic fate could be closely tied to their parents. Research, the article claims, indicates that between 33 and 40 percent of the link between a parent's and their child's income is due to genes, indicating that genetic inheritances have economic value. While intelligence quotient (IQ) might seem a likely explanation, long-time Harvard researcher Christopher Jencks argues that only a modest fraction of the economic similarity between parent and child is due to IQ, thereby suggesting that other factors (genetic and nongenetic) are more highly contributory. Traits like "ability" can be viewed as legitimate sources of economic inequality.

In America, where equal opportunity is considered a fundamental right and success seemingly achievable to anyone who works hard, factors that impair opportunity or bar success would seem ethically disturbing, if not for the call to remedial action. The issue is particularly acute if inequality results from genetic influences. If so, perspicacious ethical issues arise about what, if anything, society should do to preserve its fundamental right, that is, to ensure equal opportunity for all. Consider, for example, that nearsightedness, which is typically an inherited trait, is commonly remedied to permit equal opportunity. In rich societies, it is considered an ethical obligation to correct myopia with glasses, thereby enabling individuals to obtain the same opportunities as the normally sighted. It would be extremely unethical, in the United States, for example, to not ensure that even the poorest myopic child had glasses, not only because glasses are inexpensive but also because proper sight is essential to learning, future earnings, and socio-ability. It is, in other words, in both the child's and society's overall interest to ensure equal opportunity for optimal productivity. The real question is what, if anything, does fairness require society to do given differences in human talents, opportunities, and performance capacities? Is it fair, for example, that one will have a better shot at getting into Harvard if one's parents graduated from there? Is it fair, that some people develop better life skills because their parents nurtured them, promoting self-confidence, ambition, friendliness, sensitivity to others, and other traits linked to success? Is it fair that some people get better jobs because their parents have great connections, or live well because they inherit lots of money, or live poor because they got a lousy education in the basics of the three Rs (reading, writing, and arithmetic), or struggle to get by because their untreated childhood illness left them bereft of education and opportunity? Moreover, to the extent that any of these differences have a genetic component, what does fairness, then, require? In other words, to what extent ought biology have a role in explaining, excusing, and justifying human actions? If genes have a strong role in determining behavior, does that mean volition, or the ability to freely act or meet obligations, is therefore diminished?

Determining a reasoned position on how much genetics should matter to one's success or any number of other behaviors—or constitute an excuse or mitigating factor for bad behaviors—requires first discerning what genetics can accurately tell us (and what it can't tell us), and then determining which values are integral to defining fairness and what factors we should consider in policy making.

Behavioral Genetics

Behavioral genetics is the discipline that investigates the genetic contributions to variation in behavior. That heredity can influence human behavior is not big news. It's been known for decades, if not centuries. For example, an extra chromosome 21 causes Down syndrome, which results in mental retardation. Although scientific opinion about the hereditability of certain behaviors varies considerably, an ethical debate about the extent to which hereditability ought to factor into individuals' rights and responsibilities is relevant to any particular finding. Before assessing these issues, we must first be clear about what the science is and is not saying, so that the debate is based on accurate facts and not on prejudicial thinking or misinformation.

Background

Until fairly recently, scientists sought knowledge about the extent of genetic influence by asking, "What is the relative contribution of genes versus environment to observable traits and behaviors?" Although there was never any nature without nurture or nurture without nature, scientists nonetheless approached questions about the genetic basis of human behavior with the assumption that genes and environment exerted *independent* influences on human behavior (see Figure 4.1).[2]

This assumption was embedded within the nature versus nurture paradigm. To illustrate the fallacy in this assumption, consider explaining the hitting of a baseball as the following. Hitting a baseball with the bat is the result of the bat exerting influence A and the batter's athleticism exerting influence B, where the actual batting event (a strike, a hit, a home run) is the result of A comingling with B. We now know this assumption (that genes and environmental factors exert

Figure 4.1
Genes and Environment as Independent Influences Contributing to Phenotype

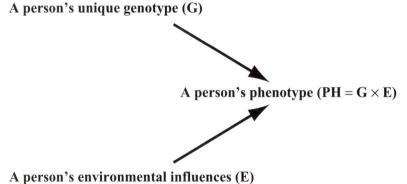

A person's unique genotype (G)

A person's phenotype (PH = G × E)

A person's environmental influences (E)

independent influences) is flawed.[3] Rather, environmental factors can initiate changes in gene expression and, conversely, genes can initiate specific responses to environmental exposures, and in particular these are complicated dynamic processes. To use the baseball analogy, the hitter's particular batting stance causes the bat itself to lengthen, shorten, or change in weight and torque, which together influence whether the ball is hit and how. In other words, knowing that genes, other biological processes, and environmental factors dynamically (reciprocally) influence phenotype, that is, behavior, means that the long-standing debate of nature versus nurture is over. For example, research shows that identical twins, who share the same genome are nonetheless on average only 60 percent identical for disease. That means that having a particular DNA sequence doesn't mean you'll develop a disease, or more significant, that you'll display a particular phenotype (or behavior).[4]

An extensive analysis of the field of behavioral genetics and the ethical concerns associated with specific behaviors is infeasible within the confines of this chapter.[5] Inasmuch as our focus is on the ethical issues arising from the potential applicability of research findings, a highly technical analysis of the science is not offered. The significance of ethics to this domain, however, is great. To that end, this chapter focuses on ethical issues associated with the hereditability of violence, particularly because this area of research is highly contentious. Not only are the public policy implications controversial, but research designs, methods, and findings are themselves hotly debated within the scientific community.

Historical Perspective

For many of us, the similarities and differences between family members and larger social groups are intriguing. Explaining why some siblings look so much alike while others look like they're not even related, or why some families are particularly athletic while other families are distinctly bookish has held a certain fascination about the possible hereditability of human traits. Curiosity about the heritability of traits, as noted in chapter 1, is at least a 6,000-year-old phenomenon.

All animal (including human) behavior is a complex phenomenon, but early scientists observed indications that at least some behaviors are hereditable. For example, many behaviors are species or breed specific. Characteristic mating rituals can distinguish birds. Consecutive generations of sheep dogs tend to display a definite herding instinct, suggesting a hereditable component to herding behavior. Ascertaining hereditable components of human behavior is arguably far more complicated, as human behavior is considered a more complex phenomenon that that of other species. People, moreover, are extremely difficult subjects for identifying hereditability for many reasons, including the comparatively long length of generations, small families, highly varied environment, and incompleteness of family records. Such difficulties, though real, clearly have not deterred scientific inquiry.

Biometrics

Sir Francis Galton (1822–1911) is credited as the first scientist to have studied heredity and human behavior. His pioneering work showed that the expression of various traits (such as height, intelligence, or certain behaviors) varies continuously from one extreme to another and results from a blending of hereditability and environmental influences. The application of statistics to biological variation to quantitatively analyze traits is now known as biometry, a field established by Galton. A brief but thorough discussion of the history of science's inquiry into human behavior can be found on the Internet.[6] As emphasized in chapter 1, advances in the science of genetics included a concomitant interest in using new knowledge to "better" humankind. Conjoining science with its application, the driving purpose of advances, arguably became prescriptive. The momentum to prevent the spread of "bad" alleles while encouraging the transmission of "good" ones developed into a moral crusade. Geneticists still concur that the value of genetics is at least in part its utility. Findings that genes contribute to violent behavior will likely continue to be relevant to our criminal justice system, including but not limited to determining responsibility, deciding appropriate punishments and restitution, and answering important questions about whether to treat the behavior, and if so, when and why.

Criminal Anthropology

The Italian physician, Cesare Lombroso (1835–1909), was the first to explore the hereditability of violence and in so doing established the field of criminal anthropology in 1870. The field encompassed both physiognomy (the art of deducing character from a person's facial and bodily features) and phrenology (studying the external characteristics of a person's skull as an indicator of their personality, abilities, or general behavioral propensities). As such, Lombroso investigated the biological basis for the differences between violent and nonviolent persons, principally through anatomy, and in doing so popularized the notion of a "natural born criminal." The notion reflects an extreme form of biological determinism, which was highly influential into the twentieth century.

In his major work, *L'Uomo Delinquente* (*Criminal Man*, 1876), Lombroso defended his assertion that crime was hereditable on the basis of evolutionary theory. Criminals are identifiable, he argued, by their anatomical signs of apishness. For example, based on the similarity of arm length, he argued, humans were genetically similar to chimpanzees.[7] Apishness, he contended, is both physical and mental, and notably "decisive" or indisputable. Lombroso defended this claim on the basis of his anatomical schemata, displaying extreme values on a normal bell-shaped curve of human physical measures (values far from human averages), from which he extrapolated behavioral traits. He further claimed that one can know a priori (independent of experience or experimental research) whether a particular individual has a *propensity* to commit crime simply by observing his or her anatomy. Humans with extreme physical measures indicating "apish" traits were considered to be potential criminals. Defining characteristics of criminality (the propensity to commit crime) included the following:

- Short or tall height
- Small head, but large face
- Small and sloping forehead
- Receding hairline
- Wrinkles on forehead and face
- Large sinus cavities or bumpy face
- Large, protruding ears
- Bumps on head, particularly the Destructiveness Center above left ear
- Protuberances (bumps) on head, in back of head and around ear
- High cheekbones
- Bushy eyebrows, tending to meet across nose
- Large eye sockets, but deep-set eyes
- Beaked nose (up or down) or flat nose
- Strong jaw line
- Fleshy lips, but thin upper lip
- Mighty incisors, abnormal teeth

- Small or weak chin
- Thin neck
- Sloping shoulders, but large chest
- Long arms
- Pointy or stubbed fingers or toes

Such bodily features formed the basis of body-type theories that were fashionable in the United States in the 1930s. Body-type theorists argued that criminals differ from noncriminals in physical respects, as proven by studies conducted in eight states. Like the scientists of the early American Eugenics Movement, these criminologists were not fringe members of society but rather respected scholars at reputable institutions, notably Ernest Hooton, at Harvard University. Hooton is credited with propounding the "truism," for example, that the Negroid forehead represented a perfect example of a criminal's forehead. Sheldon, whose work focused on juveniles, found that narrow faces, wider chests, larger waists, and bigger forearms were associated with 60 percent of delinquents and only 30 percent nondelinquents.[8] Again, the validity of deriving genotype from phenotype is believed to have been proved. Not only is the scientific basis of these assertions now known to be false, but also the logic of such reasoning is flawed. The fallacy lies in deriving genetic identity of different species on the basis of one particular phenotypic measurement. Having long arms does not predict, for example, fondness for bananas. We now know that normal variation within a population is a different biological phenomenon from the differences in average values between populations.

Despite these logical fallacies, inferring genotype from phenotype remained a valid scientific method, even beyond discrediting the scientific foundation of the American Eugenics Movement. The Eugenics Movement, which sought to eliminate certain diseases (e.g., "feeble-mindedness," assuming that being feebleminded is a medical condition) and undesirable traits (criminality) from the population, illustrates fundamental reasoning errors as well as unethical outcomes.

Davenport's Laws of Human Behavioral Inheritance

Charles Davenport (1866–1944), a pioneer of the Eugenicist Movement, the director of the Department of Experimental Evolution at the Cold Spring Harbor Laboratory, and affiliated with the Carnegie Institution of Washington, published the first book on human heredity, behavior, and eugenics in 1911. The Laboratory was also the home of the American Breeders Association. Eugenics, he claimed, is the science of the improvement of the human race by better breeding. Davenport strongly believed that unless people accept this truth and permit it to influence marriage and reproduction selection, human progress will cease. He collected "data" in efforts to provide support for this theory. Solicitations went out asking thousands of intelligent Americans, who love the truth and want to see its interests advanced, to participate in pedigree research designed to identify the inheritance patterns of a broad range of diseases, traits, and behaviors, including but not limited to the following: short or tall stature, "corpulency," special talents in music, art, literature, mechanics, invention, and mathematics; diseases such as rheumatism, multiple sclerosis, hereditary ataxia, Meniere disease, chorea of all forms, eye defects of all forms, cancer, Thomsen disease, hemophilia, exophthalmic goiter, diabetes, alkaptonuria, gout, and otosclerosis, peculiarities of hair (particularly red hair), skin, and nails; physical abnormalities such as albinism, harelip, and cleft palate; peculiarities of the teeth; peculiarities of the hands, feet, and other parts of the skeleton; feeblemindedness; and criminality. On the basis of his pedigree data, he propounded two laws of human behavioral inheritance (see Figures 4.2a and 4.2b):

Figure 4.2a

Davenport's First Law of Inheritance: Defective Parents Will Produce Only Defective Offspring

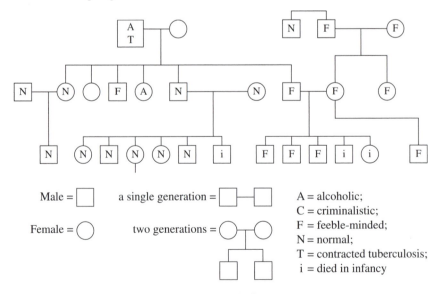

Source: Heredity in Relation to Eugenics, 1911, p. 67, fig. 31.

Figure 4.2b
Davenport's Second Law of Inheritance: Defective Offspring Are Produced Only by Defective Parents

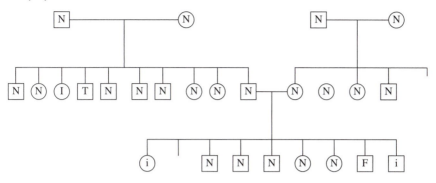

Source: *Heredity in Relation to Eugenics,* 1911, p. 70, fig. 36.

- Two mentally defective parents will produce only mentally defective offspring.
- Probably no imbecile is born except to parents who, if not mentally defective themselves, both carry mental defect in their germ plasma.

Pauperism, unlike imbecility, he believed resulted from complex causes, but genes were the primary determinant. Acknowledging that environmental causes, such as the sudden death of the father sending a family into poverty, he nevertheless held the view that poverty was caused by genetically induced mental inferiority. His proof of genetic determination lay in his derivation from pedigrees of individuals with superior mental stock. These people, he argued, did not incur poverty over several generations, which meant that, unlike the impoverished, they planned in advance for possible misfortune and therefore prevented it from ever occurring. Poverty, in other words, he claimed, resulted directly from mental inferiority.

Among several proofs for the hereditability of physical traits and behaviors was Davenport's claim to have identified the hereditability of criminality. He propounded his pedigree data as evincing the demonstrable proof that a certain kind of blood goes into the making of criminals. Furthermore, medicalizing crime by defining it as a disease, he theorized that the tendency to commit crime is found "in connection with nervous defects and disease."[9] All criminals, he argued, are a priori insane, although not all insane persons were thought to be criminals (see Figure 4.3).

While Davenport was correct that traits are neither distinctly genetic nor environmentally induced, his methods and conclusions, despite their persuasiveness at the time, are not scientifically valid. His method of

Figure 4.3
Davenport's Pedigree Proof of the Hereditability of Criminality

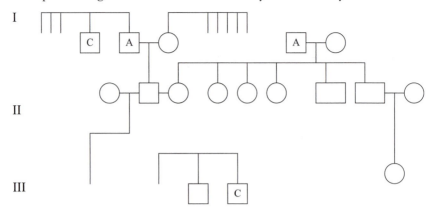

Source: *Heredity in Relation to Eugenics*, 1911, p. 86, fig. 50.

determining hereditability was essentially to deduce genotype from phenotype and was not dissimilar from Lombroso's method. He wrongly concluded that similar behaviors or traits observed within families indicated hereditability. Family histories revealing instances of criminality or aberrant behavior, for example, could be analyzed in pedigree form to reveal patterns of inheritance. His deductive argument went as follows:

- Premise: There are insane criminals.
- Premise: Criminality is observed as running in families.
- Conclusion: Therefore, the tendency toward criminality must be a hereditable trait.

Claiming that criminality is in the blood, Davenport believed his data constituted empirical proof that criminality is always fully penetrant in every generation. That implies that some individuals are criminals even before they commit crimes. Believing that an individual's behavior is controlled by their heredity, if one harbored an inherited tendency to crime, then one would inevitably commit a crime sometime in their lifetime. In other words, it was taken as impossible that an inherited a tendency to commit crime could exist but not be acted on.

Analysis and Critique of Davenport's Science

As thoughtful people, we should critically evaluate these claims for their plausibility by asking several questions. Are his pedigrees constructed objectively? Are they based on all of the available data? What criteria are espoused for determining which factors are relevant to the pedigree and which factors are not? Do we have any proof of the accuracy of the pedigrees? How good are the data? Are his "facts" selective and chosen only because they support his argument? Do the pedigrees establish causality? Do they actually prove what they purport to prove? Notably, family histories not containing instances of such behaviors or traits were excluded in the data set, and thus evidence that did not support Davenport's belief about the hereditability of criminality was discarded. Selecting only the data that support your hypothesis is obviously not credible science.

Apart from the validity of Davenport's argument are ethical issues raised by the application of his so-called research findings to society at large. Davenport argued for improving society by ridding it of all tendencies toward adversity; criminality was a case in point. Analogizing from the ethical defensibility of imprisoning the feebleminded or insane, it is just, if not morally incumbent, to potential criminals, namely, those who have inherited criminality in their blood. Imprisonment of the undesirable is then less a legal issue as it is a genetic and ethical one. Furthermore, this quest for the societal expulsion of the undesirable propelled social policy as well as laws. Involuntary sterilization of "feebleminded" persons became both lawful and seemingly moral. Indiana was the first state to enact compulsory sterilization in 1907, with other states enacting their own versions. Although many of these early laws later were declared unconstitutional, the landmark U.S. Supreme Court decision of 1927, *Buck v. Bell*, upheld Virginia's law requiring compulsory sterilization of Carey Buck, an institutionalized "mental patient" deemed to be defective due to hereditary factors. The eugenics expert Harry Laughlin, who never examined Carrie or her 7-month-old daughter (though her school teachers considered her to be very bright), supplied the "evidence" upon which the decision was based.

Logical Analysis: Scientific Method and Informal Reasoning:

Evaluating scientific claims involves analyzing the accuracy of the data provided and the extent to which it supports derived conclusions. That is, have the empirical facts presented been established by validated methods? The problem of establishing credible evidence (empirical facts) is independent, though linked, from the problem of evaluating the overall truth of a scientific claim. A key consideration in evaluating claims is whether the purported truthful conclusions are credible in that they are reasoned using logical principles. Analyzing the reasoning linking claims is important to assessing the validity of the argument espoused.

Science is not just an account of the conditions under which observed phenomena occur, it must also entail an explanation of the means by which specific effects, in certain conditions, are brought about. It involves entertaining hypotheses to explain data sets from which the hypotheses are either proven or disproved. Included in that confirmation is that under similar circumstances the hypothesis can be independently proven from a similar but different data set. Science is thus advanced one hypothesis at a time.

In establishing new knowledge, science purports to establish new laws or theories that apply without restriction. Scientific reasoning, roughly characterized, is the way in which knowledge from observed fact may be soundly extended to the unobserved. Any experimental design (intended to establish empirical facts) already involves some formulation of a specific problem, which entails the use of some concepts. An aim of a philosophical analysis is to make the underlying assumptions explicit to clarify and analyze what is implied in using them, and to determine whether such concepts fall into a kind of system capable of providing true knowledge (theories that apply without restriction to space and time).

How scientists arrive at their beliefs (how they reason) is independent from whether those beliefs actually constitute knowledge. The relationship between evidence and conclusions is such that conclusions are logical inferences from the evidence (or premises). From a logical standpoint, the truth-value of the inference depends in part on whether it is justifiable on the grounds it purports to be. Whether one has confidence in the correctness of a certain type of inference is independent of whether the inference is justified on the basis on which it is offered. Likewise, if one uses a logically incorrect method of inferring one fact from others, the facts upon which the inference is based would not constitute evidence for the particular conclusion drawn. Hence, analyzing scientific reasoning involves evaluating the way in which a conclusion (knowledge) from observed fact may be *soundly* extended.

For example, consider the following inductive argument. Most newly licensed teens are at increased risk for serious, even fatal, motor vehicle accidents and therefore as a greater insurance risk pay higher auto insurance rates. Similarly, most thrill-seekers (people who thrive on accomplishing feats such as daring skateboarding, high-risk ice climbing, tornado chasing, and so on) are greater life insurance risks and so pay greater rates. Consider, hypothetically, subsets of each cohort with inherited predispositions to alcoholism in the former, and daredevil behavior in the latter: they are even greater risks. However, consider that your cousin falls into either of those categories. While statistically (actuarially) your cousin is at greater risk, to himself and to others, it does not necessarily follow that your cousin or any single individual is a greater risk. Say your cousin is terrified of being in a car with anyone who's had the slightest amount of alcohol. Or while he participates in extreme adventure sports, he is exceedingly cautious about each and every detail. In each case, your cousin would seem to be a lower risk than the norm and less likely to be involved in an injury. Even if a tragic accident occurred, it would not necessarily, or even solely, be caused by any genetic tendency.

Some statements are not conclusively falsifiable or verifiable by observation or empirical testing. If it is reasonable to accept such statements as true then it is reasonable to accept those statements as conclusions of arguments in which the premises fail to logically entail the conclusions. Such arguments are not necessarily truth preserving or valid deductive arguments. They are traditionally referred to instead as inductive arguments. In an inductive argument, the premises are evidence for the conclusion or hypothesis. A good inductive argument is one whose premises, if true, establish the conclusion as being more likely to be true than competing conclusions. An inductive argument's conclusion is never proven absolutely without a doubt. Rather, we accept it with a degree of probability.

We evaluate the soundness of such arguments on the grounds that evidence must consist of true statements. If the statements are true, then it is reasonable to accept the inferred hypothesis as true. There are various ways in which inductive arguments can appear to be true and reasonable, when in fact they are not, because the strength of their argument lies not in the truth evidence contained in sentences but instead appeals to various extraneous rationales, referred to as fallacies.

Whether, and to what degree, science is an inductive activity is debatable and will not be resolved here. Suffice it for this purpose to note that the principle of induction is applicable to Davenport's reasoning and ultimately to his "scientific" claims. It is useful to apply these reasoning tools to these claims because doing so can serve as a model for evaluating hereditability claims reported in the media as well as in scientific literature.

Davenport starts with experimental evidence (his premises are his descriptions of his experimental results). To test the credibility of Davenport's argument, one can investigate whether his reasoning illustrates any of the aforementioned fallacies. In fact, on inspection it does.

First, Davenport assumes genetic involvement in each characteristic he investigates, failing to establish a method to distinguish genetic from environmental involvement. In other words, he uses circular reasoning, *presuming* the significant hereditability of each trait he investigates without having *proven* it. Furthermore, his terms are not adequately defined. For example, what is criminality? It is a phenomenon that can be objectively measured? Can one measure the *tendency* to commit crime? Incarceration surely can't be a measure, because it counts only those who've been caught. Has Davenport proven that criminality is biological or accepted this uncritically as being true? Or, is criminality a social construct? That is, is the definition of a particular behavior as criminal less a matter of biology than what society chooses to define as criminal?[10] As an example, consider that genital mutilation is ethically sanctioned in some parts of the world, but criminal in the United States.

Furthermore, even if his statistical correlations were valid, they do not constitute proof of causation. That is, even if feeblemindedness is associated with criminality, the association does not prove that feeblemindedness causes criminality. Consider a more current illustration of the same notion, namely, that analogous reasoning can lead to erroneous conclusions. Take the hypothesis that "African Americans are more likely to commit a crime than Asian Americans" based on the (presumed or proven) fact that a majority of inmates in prisons in this country are African American. If criminality were studied using a method for measuring heritability, the inheritance of genes controlling skin color (dark skin) would be found to be significantly correlated with incarceration. It would be incorrect to conclude that the genes producing skin pigmentation cause criminality or that melanin is a product of a criminal gene.

The standard of proof, or validity, for Davenport lies in the perceived consistency of a behavior within a family. However, consistent perceptions do not constitute scientific fact. One family may overreport a particular characteristic while another family may underreport the very same phenomena. Even so-called objective measurements can be subject to similar criticism, for inaccuracies may be due to individual error in measurement, the use of noncalibrated measuring equipment, improper study design, and many other factors. Furthermore, such reports are based on memory, a factor inherently biased and variable. Though recollection is an accepted data collection technique, various biases must be taken into account if recollection is used as a valid method to obtain quality data. Such biases are not acknowledged by Davenport. Another key concern is the reliability of the data. Are the conclusions reproducible? Would another investigator reach the same conclusions? Innumerable value-laden assumptions taint his data-derived conclusions. Presuming that an individual's chosen occupation indicates their highest level of competence, Davenport, for example, concludes that employment as a stenographer signified low mental ability and not possibly the chosen profession of a genius. Yet, the possibility that one took a menial job for any other reason, such as the only available means of supporting a family, is not considered.

Ethical Analysis

Apart from the scientific validity of early behavioral genetics, is a core ethical question about when (if ever) scientific knowledge should be socially applied. The ability to use such knowledge does not entail that we ought to. Applying the ethical theories discussed in chapter 2 to this question, can illustrate a broad range of ethical arguments, and different underlying rationales, in favor and against. For example, should we apply the new knowledge to benefit the majority, that is, the Utilitarian argument, and if so, what is a benefit? How can the concept of benefiting the majority be defined in this instance? Ought thresholds for applicability vary depending on certain external factors, like overcrowding of the criminally insane in psychiatric institutions? Ethical concerns are nearly limitless. The central point, though, is that ethical issues about the social applications of new scientific knowledge remain vital concerns.

Dawn of the New Genetic Era

Defining the hereditability of criminality on the basis of physical traits continued well into the 1960s when the super-male syndrome (XYY) was associated with criminality. At the time, new biological methods purported to offer a way to link "abnormal" behavior to chromosomes. A 1965 study found chromosomal aberration in a small sample of mentally retarded male criminals and the genotype XYY was postulated as a cause of excessive aggressive behavior, including criminality. Subsequent studies correlated the genotype with reading problems and other learning disabilities. Interest in identifying such children early on to provide educational interventions emerged. One study designed to identify newborns was awarded federal funds but later was halted on the grounds that the research was inherently unethical. First, identified individuals (males with XYY genotypes) would likely be stigmatized for life, because the availability of their genotype would indicate an abnormality. Without the genotype identified, they would be indistinguishable from other males. Second, the study could not be objectively conducted, because parents who were informed that their son was XYY in all likelihood would overreport instances of aggressive behavior. Moreover, parental perceptions of aggressive behavior were hardly an accurate tool for measuring cause and effect of the XYY genotype. Such methodological and ethical problems are not too dissimilar from Davenport's errors. An overarching ethical question is whether it is ever morally acceptable to do predictive experiments on random populations of normal individuals, since the very fact of participation is likely to significantly change their lives and therefore affect the outcome.

Despite methodological, ethical, and policy problems, the quest to find a genetic basis for aggression continues. Notably, this drive exists not only within the scientific community, but also among families desperate for answers to apparent familial tendencies.

In 1978, a Dutch woman sought the help of a clinical genetics researcher, H. G. Brunner, to explain the apparent tendency of males in her family toward unprovoked aggressive outbursts. She came with family history compiled over 30 years by an unaffected maternal great uncle. The history described nine male relatives, all of whom displayed the aberrant behavior. In addition to the uncle's observations, she claimed that five more males in the family "developed the disorder." Bruner undertook her request. His pedigree analysis revealed a single sex-linked phenotype (e.g., the gene expression only affected males). Results indicated that a mutation in the gene for MAOA (monaminoxidase A) might underlie aggressive and occasionally violent behavior displayed by certain males over generations in a single family. Biochemical analysis of urine samples showed highly elevated amounts of monoamine (dopamine, epinephrine, and serotonin) neurotransmitters, indicating that MAOA, the enzyme that breaks down these neurotransmitters, was likely malfunctioning in such individuals. Unmetabolized neurotransmitters might in effect put the person in "overdrive," which would make coping with stress more difficult.[11] Most males had an IQ of 85 and exhibited a wide range of violent behaviors, including rape, violent unprovoked attacks, attempted murder, and arson. The preponderance of familial violent behavior over generations suggested the possibility of a genetic contribution, if not an explanation.

Newer evidence supporting a genetic influence to aggressive behavior continues to emerge.[12] Studies have shown that low concentrations of serotonin metabolite 5-HIAA (5-hydroxyindoleacetic acid) in cerebrospinal fluid correlate with impulsive behavior.[13] Based on DNA linkage studies, Brunner claimed to have identified the causative gene. Family, twin, and adoption studies added support to a biological basis for anti-social behavior.[14]

Knowing that it is impossible to isolate genes from the environmental contexts in which they express themselves to determine which factor controls what behaviors or traits, geneticists now use the following algebraic equations to describe the origins of variability:

$$Vp = Vg + Ve \text{ and that } Vp = VgXVe^{15}$$

Total phenotypic variability (Vp) is made of up of variation in genotypes (Vg) and variation in environments (Ve).[16] We now know more about how genes express themselves (phenotype) and the variation that occurs in expression (phenotypes), such as eye color or dominant hand. Each is a quantitative trait whose expression is attributable to the inheritance of several genes (polygenic inheritance), although the exact mode of inheritance is presently unknown. In addition, numerous other equations to assess polygenic effects are used in combination with the equation above, lest this simplified discussion appear as grossly reductionistic approach.

Some researchers have argued, for example, that biochemical levels of serotonin and testosterone are correlated with violent behavior. Correlation is not, as noted above, causation. To illustrate this difference, consider the following simplistic argument. Suppose that violence (A) is associated with eating large quantities of pizza (C). Suppose low serotonin levels (B) are similarly correlated to pizza intake (C). The question is whether it is reasonable to infer that violence (A) is associated with serotonin (B) on the basis that it is associated with eating large quantities of pizza (C). The argument takes the form: A is associated with C; B is associated with C; therefore by analogy (or substitution) A is associated with B. The question is whether it is reasonable to infer the association of A with B. Even if it is generally true that A is correlated with B, there is the question of whether such a correlation is true in every case or in any particular case. In other words, is it true that the serotonin level (B) of a particular individual (Harry) caused him to commit a specific act of violence (the rape and murder of a woman)? Even if Harry committed a violent act and his serotonin level was in the range shown to be correlated with violent behavior, there may well be a myriad of other biochemical, volitional, and situational causes for his action. That is, what is generally true in a particular case is not necessarily true in every case. Unless you ask the question of whether some factor disassociated with biochemical factors could be responsible for violent behavior, you haven't proven that biochemistry is a meaningful cause of the behavior.

Moreover, in assessing the truth-value of such claims, you would want to know how the biochemical findings of the studied population compare with those levels in the general population. In other words, generalizing hypotheses from findings in controlled research settings to the likely truths about behaviors outside of a research context involves assessing the argument and not merely determining if the science is true.

Ethics of What Is Now Known

While much of the research in this area is increasingly sophisticated and challenging for anyone untrained in the methods used, philosophical issues remain about whether behavior is wholly reducible and primarily understood in biological terms, especially when assessing scientific claims of causality and the implications for exercising free will and self-determination. Despite these concerns, the quest for a genetic basis of violence continues (with studies purporting positive evidence) to strive to demonstrate genetic influences. Recent research suggests the genetic basis for male aggression.[17] Another study examining 1500 pairs of Swedish and British twins found the male inheritance of aggressive behavior.[18] Researchers concluded that boy bullies are more likely to have inherited their behavior through their genes than girl bullies. In 2001, a study of 492 twin pairs found a strong genetic component to aggression and genetic contributions to anxiety, depression, and inattentiveness.[19] The study asked one parent of each twin to fill out the Child Behavior Checklist and compared monozygotic with dizygotic twins[20] as a way to distinguish between environmental and hereditary influences. In September 2002, the article "Mean Genes: Why Some People are Jerks," reported that genes can account for 30 to 60 percent of observed variability in personality, and that abusive behavior is associated with genes, based on a large Minnesota Twin study.[21] In May 2003, researchers reported identifying a mutation in the gene Nr2e1 for a group of brain receptors preserved across numerous species is linked to violent behavior in mice.[22] Additional studies found that the absence of the PET-1 gene in mice (also found in humans) seemed to cause suffering,[23] and scientists concluded that "people who are over-aggressive or excessively anxious may be missing this gene." Evidence suggested that mice lacking the gene suffered.

While this account of findings is oversimplified, it is easy to overinterpret their significance and lose sight of the fact that behavioral traits are mutifactorial (involving many factors both genetic and environmental), often polygenetic (involving more than one gene), and influenced by penetrance (the extent to which a particular genotype is expressed). When a genotype is uniformly expressed, it is said to be fully penetrant, though the gene expression can be variable. Incomplete penetrance refers to the presence of a specific genetic allele that may not be expressed clinically. Such a gene can be transmitted with different expression in the subsequent generation. Variable expression makes it difficult to assign causative roles to genes and, thus, it is difficult to link genes meaningfully to psychosocial problems.

Crime and antisocial behavior are widely accepted as resulting from factors, perhaps polygenetic influences, and most definitely environmental influences.[24] Notably, most contemporary scientists who believe that human behaviors result from genetic factors also believe that changes in environmental conditions (such as education and training) can mitigate genetic predispositions. For example, researchers who believe that boys have right brain versus left brain advantages over girls that predispose them to a greater facility with mathematics nonetheless believe that, with training to compensate for this difference, girls are capable of attaining the same level of performance. That is not to say that difficulties in defining and measuring behavioral traits do not currently exist, for they do. While many studies have difficulty in withstanding rigorous analysis, it is increasingly likely that genes have a role in behavioral traits.

The social utility of these findings remains controversial. Despite the high level of interest in the genetic basis of violence, research is far too preliminary to be applied to the real world quite yet, and any application raises ethical concerns. It is probable that in the coming decades research results will be sufficiently compelling to be applied in some fashion. In the future, it might be possible to identify such predispositions through DNA-based testing, though the utility of testing is questionable in light of sensitivity and specificity limitations. No test is ever 100 percent accurate. Some mutations may not be picked up by the test, hence, a probability of false-negative results. A positive test result may not necessarily predict inevitable violent behavior. Moreover, the presence of a positive test and a violent act does not in and of itself prove a causal connection, and violence might have occurred for reasons independent of a predictive test result. Nonetheless, if a genetic test is developed, the question of intervention is not far behind. While it is impossible to predict whether medicines will be developed to treat certain traits, any tendency to widen the existing diagnostic categories is likely to encourage treatment for any display of aberrant behavior, expand the realm of what is considered aberrant, and reduce tolerance of individual idiosyncrasy. Clearly, compelling scientific data are independent of ethical questions about the justifiability of using detection and treatment methods. If testing and interventions were cost-effective, they would present government with a compelling alternative to paying for the apprehension, legal proceedings, and imprisonment of violent criminals. Moreover, the availability of such tests might be an incentive for couples to undergo prenatal testing to avoid the birth of children with predispositions to violence. Employers similarly might want pre-employment testing to avoid hiring workers who test positive for the gene(s).

Parallel ethical issues pertain to questions of treatment. Perhaps treatments for adults capable of giving voluntary consent would be ethically acceptable, but what about individuals deemed incapable of providing consent, such as young children? Should parents be free to treat their children who have tested positive but do not exhibit "behavioral" symptoms to prevent criminal behavior? Would a morally acceptable decision rest on knowing short- or long-term effects of medication or other interventions, or would the indication for treatment transcend such pragmatic concerns? What if treatment long after it was administered is found to have late onset adverse effects that are conceivably worse than the original propensity to violence? What if the intervention was not simply a pill or shot, but rather **germ-line gene therapy** designed to benefit not only the individual but also his or her future offspring? Who might have access to and who might be barred from treatments? Would insurers require such treatments to reduce health care costs? Would Medicaid and Medicare require enrollees to be treated as a condition of enrollment status? What does justice require? These are a few of the many and varied ethical concerns that we might face in the future.

Gene therapy (particularly germ-line gene therapy) represents the next logical step. Although given present safety concerns[25] and functional success—the possibility that the inserted gene could be expressed in the wrong place at the wrong time, such as interrupting normal gene function or precipitating disease by turning on or off certain genes (e.g., turn on an oncogene or turn off a DNA repair gene) is a current obstacle—it is not yet feasible. Nonetheless, ethical concerns arising from the possibility are significant enough for the United Nations Educational, Scientific, and Cultural Organization (UNESCO) to have stated in its Universal Declaration on the Human Genome and Human Rights, in Article 5, that

> research, treatment or diagnosis affecting an individual's genome shall be undertaken only after rigorous and prior assessment of the potential risks

and benefits pertaining thereto and in accordance with any other requirement of national law.[26]

International consensus favors the avoidance of germ-line gene therapy in general, particularly for violent behavior and other socially constructed traits.

Some have argued that any research into the genetic basis of human behavior is an outright affront to human dignity. Humans, the argument goes, are clearly more than the sum total of their genome. For those who believe that God is responsible for human life and purpose, such genetic pursuits may be considered immoral. How can we (according to proponents), the divine creatures that we are, be reduced to being merely biologically determined or genetically programmed? Religion and science may always make strange bedfellows, yet neither scientific evidence nor religious beliefs will resolve such bones of contention. Of ethical relevance is whether we as individuals, as a society, or as a world have the right to pursue knowledge of the genetic influences of human behavior for its own sake or for the sake of improving, in some benevolent way, humankind.

Fiscal considerations come into play in discussions of the feasibility and desirability of applying scientific knowledge. Consider research that indicates that it is men, and not women, who commit the vast majority of all crimes. The evidence is so statistically strong that it is fair to conclude with near certainty that men are far more likely than women to commit the majority of all crimes in the future. In 1991, for example, 93 percent of the first-time offenders arrested for a violent crime were male as opposed to 7 percent female. During that year, roughly 88 percent of all crimes were committed by men.[27] The disparity remains. As of 2001, for example, far more men than women were imprisoned.[28] The most recent statistics from the Bureau of Justice show that the lifetime likelihood of going to state or federal prison is far greater for men (11.3 percent) than for women (1.8 percent).[29] In the early 1990s, one prison cell cost $100,000 to construct, not including interest paid on bonds raised for prison construction. Prison construction costs have risen sharply since then, as have the costs of maintaining prisoners. By 2000, New York State reported that their costs had risen 148 percent since 1983.[30] Furthermore, the financial burdens of crime are not restricted to prison costs. There are replacement costs from property loss, medical and other therapeutic costs of caring for the victims, state costs resulting from disabled victims, and costs involved with failure to pay child support, as well as the cost incurred by loss of what would otherwise be more productive societal contributions. Data indicate that men are clearly a significant cost burden to society, at least in the United States. Applying fiscal analyses to men, June Stephenson argued that men simply are not cost-effective.[31] To assess the overall cost-effectiveness of their existence, however, we would need to determine the societal benefits versus costs, where benefits and burdens are measured quantitatively but also qualitatively. Qualitative measures might indicate, for example, that despite their financial burden, males confer social benefits, say, their superior physical strength has led to societal benefits that woman would not have been able to achieve on their own, that balance, if not outweigh, their burdens.

James Q. Wilson, one of several researchers studying crime and genetics, claimed that criminality requires a particular personality trait in a particular environment. In *Crime and Heredity*,[32] he argues that males have distinctly different personality traits than female that make them susceptible to crime. Men and women in the same environment (one that rewards for criminal behavior) might behave differently, with men and not women committing crime. The "boys will be boys" cliché essentially means that males behave the way they do because "it is in their nature" and they are biologically wired to be aggressive. The cliché continues to mitigate and even excuse bad behavior, as if the perpetrating male trait is either resistant to moral development, at least to the same moral development that girls undergo during development.

We do know that moral development is tied to biology, for as children grow physically, they also gain an increasing awareness of themselves as social beings. Living in a civic society that has certain norms of right and wrong confers certain rights and responsibilities. As a society, we decide which rights and privileges are appropriately given at what stages of moral development. For example, 16 is generally the age at which individuals can give consent to medical treatments. States set age thresholds for driving, marriage, and statutory rape. Other rights and responsibilities are conferred socially rather than legislated, like the age at which an individual can commit certain acts of omission without censure or liability. Witnessing a violent crime but doing nothing about it, for example, does not amount to violating the law, though it is generally viewed as morally reprehensible. Arguably, the highest stage of moral development is exercising one's own conscience, which is doing the right thing even if you don't stand to benefit. The reward for doing the right thing is feeling good about yourself and your ability to do the right thing in the face of temptations to do otherwise. The punishment for doing the wrong thing is suffering the weight of your conscience. The relevant question could then become, to what extent is a person's conscience biologically determined?

Moral development, though tied to biology, is not reducible to it. Yet the biological contribution to moral aptitude is clearly at issue when we discuss the genetic basis of "bad behavior." Though it may be common parlance to speak metaphorically about our genes as causing us to behave in certain ways, genes do not predetermine our will to act or our behavior. While genetic factors may well influence behaviors, and reductionist or functionalist accounts have appeal, genes do not fully determine an individual's free will or autonomy. Although they augment our understanding, as yet they provide no reason to radically alter our common-sense understanding of human choice (free will or autonomy). What proportion of free will is genetically determined versus environmentally determined or a function of gene-environmental interaction is debatable, and rests not only on scientific evidence but on moral grounds.

Analysts have pondered whether genetic information associated with behavioral traits will alter our notion of autonomy, and if so, whether transformed notions ought to translate into public policy and law. For example, could it ever be morally acceptable to compromise the autonomy of an individual with identified predispositions to aggression even if they have not exhibited aggression publicly or privately? When, for instance, is "normal" anger no longer anger but rather aggression? While the research into behavioral genetics has the potential to benefit us by deepening our knowledge of humans, the genetic contribution is but one among other influences. The ethical issues are nearly endless if not irresolvable. Scientific researchers have turned to policy makers and ethicists to wrestle with these issues, and policy makers have turned to the citizenry to participate in debate and policy setting. Arguably, each of us has an obligation to inform ourselves about the science and its ethical underpinnings so we can act responsibly in helping to set the policy agenda to ensure that every person is treated with respect and dignity. Conscious awareness of the issues and the application of ethical theories and principles to issues can help to clarify arguments, focus thinking, and assess the moral justification of proposed applications that go well beyond a gut level feeling about the right thing to do.

Notes

1. "Is Success in Your Genes?" *Wall Street Journal*, May 26, 2005.

2. Institute of Medicine Report, "Genes, Behavior and the Social Environment: Moving Beyond the Nature/Nurture Debate," http://www.iom.edu/CMS/3740/24591/36574.aspx.

3. The notion that nature and nurture are independent of one another and exert separate influences (nature exerting genetic contribution and nurture exerting environmental influences) that combine to produce a phenotype is now recognized as flawed. Mange, E. J., and Mange, A. P., *Basic Human Genetics* (Sunderland, MA: Sinauer Associates, Inc., 1994).

4. Sluyter, F., Arseneault, L., Moffitt, T. E., Veenema, A. H., de Boer, S., and Koolhaas, J. M., "Toward an Animal Model for Antisocial Behavior: Parallels Between Mice and Humans," *Behavioural Genetics* 33, no. 5 (September 2003): 563–74; Alper, J., et al., eds., *The Double-Edged Helix: Social Implications of Genetics in a Diverse Society* (Baltimore: Johns Hopkins University Press, 2002).

5. See Public Broadcasting System, *A Scientific Odyssey: Human Behavior*, http://www.pbs.org/wgbh/aso/thenandnow/humbeh.html; http://www.pbs.org/inthe balance/archives/ourgenes/genes_on_trial/genes.resources.html.

6. *Newsweek*, June 4, 2007, 48–50.

7. Lombroso, Caesar, *The Criminal Man* (1876; reprint, Durham, NC: Duke University Press, 2006); Gibson, Mary, *Born to Crime* (Westport, CT: Praeger, 2002); Gould, S. J., *Mismeasure of Man* (New York: W. W. Norton, 1981, 1996), 124.

8. Sheldon, William, *Varieties of Delinquent Youth: An Introduction to Constitutional Psychiatry* (New York: Harper, 1949).

9. Davenport, Charles B., "The Inheritance of Family Traits," in *Heredity in Relation to Eugenics* (New York: Holt, 1911), 82.

10. See Wasserman, David, "Science and Social Harm: Genetic Research into Crime and Violence," *Report from the Institute for Philosophy and Public Policy* 15, no. 1 (Winter 1995): 14–49 for commentary.

11. Brunner, H. G., Nelen, M., Breakefield, X. O., Ropers, H. H., van Oost, B. A., "Abnormal Behavior Associated with a Point Mutation in the Structural Gene for Monoamine Oxidase A," *Science* 262, no. 5133 (October 22, 1993): 578–80.

12. Brunner, H. G., "MAOA Deficiency and Abnormal Behavior: Perspectives on an Association," *Ciba Foundation Symposium* 194 (1996): 155–64, discussion 164–67; Craig, I. W., "The Role of Monoamine Oxidase A, MAOA, in the Etiology of Antisocial Behavior: The Importance of Gene-Environment Interactions," and Lovell-Badge, R., "Aggressive Behavior: Contributions from Genes on the Y Chromosome," *Novartis Foundation Symposium* 268 (2005): 20–33, discussion 33–41, 96–99; *Novartis Foundation Symposium* 268 (2005): 227–37, discussion 237–41, 242–53.

13. Kruesi, M. J., Rapoport, J. L., Hamburger, S., Hibbs, E., Potter, W. Z., Lenane, M., and Brown, G. L., "Cerebrospinal Fluid Monoamine Metabolites, Aggression, and Impulsivity in Disruptive Behavior Disorders of Children and Adolescents," *Archives of General Psychiatry* 47, no. 5 (May 1990): 419–26; Brown, G. L., and Linnoila, M. I., "CSF Serotonin Metabolite (5-HIAA) Studies in Depression, Impulsivity, and Violence," *Journal of Clinical Psychiatry*

51, suppl. (April 1990): 31–41, discussion 42–43; Mehlman, P. T., Higley, J. D., Faucher, I., Lilly, A. A., Taub, D. M., Vickers, J., Suomi, S. J., and Linnoila, M., "Low CSF 5-HIAA Concentrations and Severe Aggression and Impaired Impulse Control in Nonhuman Primates," *American Journal of Psychiatry* 151, no. 10 (October 1994): 1485–91.

14. Nelson, R. J., and Chiavegatto, S., "Molecular Basis of Aggression," *Trends in Neuroscience* 24, no. 12 (December 2001): 713–19; Mong, J. A., and Pfaff, D. W., "Hormonal and Genetic Influences Underlying Arousal as It Drives Sex and Aggression in Animal and Human Brains," *Neurobiology of Aging* 24, suppl. 1 (May–June 2003): S83–88, discussion S91–113.

15. The original equation of Vp = Vg + Ve simplifies and even mischaracterizes the now known to be more dynamic interaction of genes and environment. This second equation recognizes this dynamism. We now know that P = G x E (Gv x E; G x Ev; or Gv x Ev). Gv x Ev represents the phenotypic variability in siblings. G x Ev represents identical twins raised in different environments. Gv x Ev represents variability in large populations.

16. Mange, E. J., and Mange, A. P., *Basic Human Genetics* (Sunderland, MA: Sinauer Associates, Inc., 1994).

17. Le Roy, I., Mortaud, S., Tordjman, S., Donsez-Darcel, E., Carlier, M., Degrelle, H., and Roubertoux, P. L., "Genetic Correlation Between Steroid Sulfatase Concentration and Initiation of Attack Behavior in Mice," *Behavior Genetics* 29, no. 2 (March 1999): 131–36; erratum in *Behavior Genetics* 30, no. 3 (May 2000): 249; "Study: Abuse and Genetics = Aggression," August 1, 2002, http://www.cbsnews.com/stories/2002/08/01/health/printable5174241.shtml; see also Moffitt, T., et al., "Males on the Life-Course-Persistent and Adolescence-Limited Antisocial Pathways: Follow-up at Age 26 Years," *Development and Psychopathology* 14, no. 1 (Winter 2002): 179–207.

18. Eley, T. C., Lichtenstein, P., and Stevenson, J., "Sex Differences in the Etiology of Aggressive and Non-aggressive Antisocial Behavior: Results from Two Twin Studies," *Child Development* 70, no. 1 (January–February 1999): 155–68.

19. "Linking Brain Dysfunction to Disordered/Criminal/Psychopathic Behavior," *Crime Times* 7, no. 1 (2001): 7; see also Hudziak, J., et al., "A Twin Study of Inattentive, Aggressive and Anxious/Depressed Behaviors," *Journal of the American Academy of Child and Adolescent Psychiatry* 39 (2000): 469–76.

20. Monozygotic twins are commonly known as identical twins. This type of twins derives from a single egg fertilized by one sperm that divides into two separate cell masses within the first two weeks of development. Monozygotic twins are the same sex and their genotypes are identical except for possible somatic mutations. Somatic mutations are mutations that occur in somatic cells. Somatic cells are the cells other than germ and gamete cells.

21. Patterson, Michelle, "Mean Genes: Why Some People are Jerks," http://www.expressnews,ualberta.ca/expressnews/articles/printer.cfm?p_ID=2927.

22. University of British Columbia, "UBC Researcher Finds Genetic Link to Aggression," Press Release, May 6, 2002; see also Young, K. A., et al.,

"Fierce: A New Mouse Deletion of Nr2e1; Violent Behavior and Ocular Abnormalities Are Background-Dependent," *Behavioural Brain Research* 132, no. 2 (May 14, 2002): 145–58.

23. Hendricks, T. J., et al., "PET-1 ETS Gene Plays a Critical Role in 5-HT Neuron Development and Is Required for Normal Anxiety-Like and Aggressive Behavior," *Neuron* 37, no. 2 (January 23, 2003): 233–47; http://news.bbc.co.uk/2/hi/health/2692299.stm.

24. Nuffield Bioethics Council on Bioethics Report 2001, "Genetics and Human Behavior," http://www.nuffieldbioethics.org/publications/geneticsandhb/rep0000001105.asp.

25. Report from the Committee on the Ethics of Gene Therapy, Sir Cecil Clothier Chairman, British House of Commons, 1992, http://www.bopcris.ac.uk/imgall/ref23308_1_1.html.

26. UNESCO, Universal Declaration on the Human Genome and Human Rights, November 11, 1997, http://portal.unesco.org/en/ev.php-URL_ID=13177&URL_DO=DO_TOPIC&URL_SECTION=201.html.

27. Report from the Committee on the Ethics of Gene Therapy, Sir Cecil Clothier Chairman, British House of Commons, 1992, http://www.bopcris.ac.uk/imgall/ref23308_1_1.html.

28. Higher for Black males (16.6 percent) and Hispanic males (7.7 percent) than for white males (2.6 percent); higher for black females (1.7 percent) and Hispanic females (0.7 percent) than white females (0.3 percent), http://www.ojp.usdoj.gov/bjs/crimoff.htm.

29. http://www.ojp.usdoj.gov/bjs/cvict.htm.

30. http://www.ncpa.org/ea/eaja92/eaja92h.html.

31. Stephenson, J., *Men Are Not Cost-Effective: Male Crime in America* (New York: HarperPerennial, 1995).

32. Wilson, J. Q., and Hernstein, R., *Crime and Human Nature: The Definitive Study of the Causes of Crime* (New York: Free Press, 1985).

CHAPTER 5

To Do No Harm: From adenine (A), cytosine (C), guanine (G), and thymine (T) to Molecular Medicine

[I]t would be good if I could start patients off on the right dose of medicine—avoid the ill effects, the failed response, the expense and the inconvenience for doctor and patient; nothing is more dispiriting than giving medication with confidence and conviction, and then having to fiddle around with dosing and blood testing and sometimes starting all over again, sometimes more than once.

—Anonymous primary care physician (2003)

If it were not for the great variability among individuals, medicine might as well be a science and not an art.

—Sir William Osler (1892)

Medicines, like nutrients, are metabolized differently by different people. In 2000, physicians prescribed the normal pediatric dosage of Prozac to a 9-year-old boy, one of several drugs he took to treat attention deficit disorder, obsessive compulsive disorder, and Tourette syndrome. Shortly after treatment began, the boy suddenly had a heart attack and died. His autopsy revealed a Prozac concentration at several levels higher than overdose indication criteria and the medical examiner determined that the boy's death was caused by an overdose of Prozac. What was *not* known at the time of treatment was that the boy had a variation in the P450 enzymes that prevented him from metabolizing the active ingredient in Prozac. In other words, his genotype indicated that he was a slow metabolizer, which meant that the standard dosage would quickly reach toxic levels in his blood stream. Had genetic testing been used before administering the medicine, it would have informed his physician that the standard pediatric dose would be neither safe nor effective. The doctor could have then prescribed either a much lower dose or a different medication, possibly one in a different therapeutic class, that would have been safe and effective.[1] This true story illustrates not only the serious problem with our one-size-fits-all trial-and-error dosing practice, but also implies the numerous ethical issues (including but not limited to physician duty to prevent harm, if not liability for failure to do so, concerns about DNA base testing of children, an individual's right not to be tested, etc.) involved in determining the appropriate use of pharmacogenomics, or personalized medicine.

While the ability to predict drug safety and efficacy enables personalized prescribing, the application of clinical pharmacogenomics raises a number of ethical concerns focused not only on the personal nature of genetic information, but also on its familial and communal nature, raising concerns for individuals and for groups of individuals and ultimately about global benefit sharing. For example, knowing the dangers of trial-and-error prescribing raises an important ethical question: How do physicians who, under their Hippocratic Oath are obligated to do no harm, fulfill this duty when serious harm (e.g., from adverse drug reactions [ADRs]) will befall some patients? Although pharmacogenomic-based prescribing exists,[2] it is far from the standard of care. It behooves us to consider the panoply of ethical issues surrounding pharmacogenomic adoption to ensure that adoption—be it public health surveillance, clinical care standards, or public policy—meets ethical requirements, including but not limited to equity and justice. Not all of the ethical issues discussed here are unique to pharmacogenetics and pharmacogenomics.

Background

Pythagoras in sixth century B.C. advised individuals to avoid fava beans, because he had observed that some people became sick upon eating them. His advice proved sound when twentieth-century scientists discovered that there is biological explanation for this difference, namely, a genetic deficiency that in some individuals causes acute hemolytic anemia (a serious red blood cell disorder) when they ingest the cooked beans.

Around the mid-1800s physiological chemists observed that most medicines were excreted in different forms from which they were ingested, demonstrating that they had been chemically altered in the process of metabolizing them for therapeutic effect. Archibald Garrod, and other researchers at the time, discovered that an individual's enzymes play a significant role in detoxifying foreign substances, be they medicines, foods or other substances, with the detox mechanism functioning well in some people and not at all in others. This work spurred future findings that formed the basis of early pharmacogenetics (the phenotype-based study of identifying variations in genes that influence drug metabolism)—namely, that enzymes mediate certain metabolic pathways, particularly those in the endoplasmic reticulum of the liver, that biotransform foreign chemicals (in some cases poisons) into harmless by-products.[3]

Genetic research at the time determined the hereditability of certain enzymatic functions, particularly the inherited specificity of taste and smell response to chemicals. New technologies in the 1950s enabled researchers to separate similar proteins to study patterns of drug metabolites in different individuals, and thus begin to link the metabolism of foreign chemicals to the genetic involvement in an individual's response to drugs. Among the first such studies was research demonstrating hereditary variance in serum cholinesterase, the normal metabolic product of succinylcholine, a muscle relaxant, used in sugary foods and electroshock therapy. Typically, succinylcholine is broken down into inactive products and serum cholinesterase, but research showed that some individuals and their relatives, all of whom had an abnormal enzyme, experienced an adverse reaction caused by the slow metabolism of cholinesterase. Hereditary variation in responses to isoniazid, an effective agent against tuberculosis, causes certain individuals to display peripheral neurotoxicity on a therapeutic dose. Their problem was that they were slow acetylators.[4] Subsequent molecular studies have confirmed this finding of "slow acetylation." We now know that mutations of N-acetyltransferases are catalytically impaired and cause impaired acetylation.

Alf Alving, around the same time, discovered that a large fraction of African-American males (and people of Mediterranean descent) have a specific hereditary red blood cell enzyme glucose-6-phosphate dehydrogenase (G6PD) that causes red blood cell destruction when antimalarials and sulfa antibiotics are ingested. Early research showed that 10 percent of African-American men serving in the Korean War became anemic after ingesting an antimalarial drug, which rarely, if ever caused problems for Caucasian soldiers. The anemic reaction was determined to be caused by a variation of the G6PD gene, and subsequently the variation was found to be common among people of African descent but not among Caucasians. It was later discovered that the nonvariant form of the gene makes an enzyme that protects red blood cells against certain chemicals. Lacking that protective effect, those with the variant form are vulnerable to deleterious hemolysis. Deficiencies in enzymes can explain adverse reactions to drugs and those variations could be inherited.

Pharmacogenetics and Person-to-Person Variation in Drug Response

Many mark the true beginning of pharmacogenetics with Arno Mutulsky's seminal 1957 *Journal of the American Medical Association* paper entitled "Drug Reactions, Enzymes and Biochemical Genetics" about person-to-person drug response differences. Individuals display significant differences in response to therapeutic agents, in part because of differences in hereditary material that is transmissible from one generation to the next. Hereditable variation can manifest itself in various ways, including acute toxicity, delayed toxicity, drug or disease resistance, unwanted drug or chemical interactions, and susceptibility to unsafe or inefficacious effects.

For medicines to work in the body, they must activate a specific ingredient from its otherwise inactive state. For example, codeine is an inactive analgesic until converted into the active substance morphine, which provides the pain relief. Specific liver enzymes and proteins are required to properly convert codeine to its active state. These biochemicals belong to a family of enzymes called the cytochrome P450 enzymes (CYP450) and are primarily involved in drug metabolism. This family of enzymes is a function of not a single gene but rather 60 genes. The P450 enzymes are involved in metabolizing roughly one-fourth of all prescription drugs. Variations in these enzymes can produce enormous differences in patient reactions to medicines.

Several types of CYP450 enzymes activate or remove therapeutic agents from the body. Other enzymes in this family thought to cause drug interactions and adverse reactions include CYP3A4, CYP2C9, CYP2C19, and CYP2D6. CYP2D6 is one of the most important of these genes and is involved in metabolizing dozens of commonly prescribed drugs, such as antiarrhythmics, antidepressants, beta-blockers, analgesics, and antipsychotics. Tiny differences in one of these enzymes can cause too much of an active ingredient to be produced. People with these variants are referred to as ultra-rapid metabolizers. Other differences cause either too little of the active ingredient to be produced or no conversion to the active ingredient at all. People with these variants are called slow metabolizers. Ultra-rapid metabolizers break down and dispel a drug so fully that the drug is eliminated without having generated the intended therapeutic effect. Conversely, slow metabolizers break down a drug so incompletely that quantities of the active ingredient remain in the system and build to toxic levels, causing severe, even life-threatening effects.

To understand what this means, consider a patient who is genetically predisposed to be a poor metabolizer of a particular drug or class of drugs, say codeine. Some 5 percent of the general population lack an enzyme needed to convert codeine into its active form of morphine and so receive no therapeutic benefit (or pain relief) at normal or high doses. An estimated 1 percent of the U.S. population are poor metabolizers, meaning their bodies either dispel the medicine without obtaining the intended therapeutic effect or cannot properly break down the ingredients and, as a result, concentrations of the medicine is retained in the body at toxic levels. Five to 10 percent of the Caucasian and African-American population are poor metabolizers and few Asian Americans have the poor metabolizing allele. The ultra-rapid metabolizing allele is common among Ethiopian and Saudi Arabian populations. Ten to 20 percent of the Japanese population has the CYP2C19 variant that is ineffective and roughly 4 percent of Caucasians have this variant.

Pharmacogenetics can identify people who are ultra-rapid metabolizers and slow metabolizers, thus enabling prescribers to predict whether a medicine will be safe and effective. The present trial-and-error nature of prescribing functionally means that lots of drugs do not work for lots of patients. When medicines don't work, at best they are ineffective and at worst they are seriously unsafe, causing ADRs that sometimes lead to death. We pay a large price for medication failures, both individually and socially. To scale the magnitude of this problem, consider that in 1998 a total of 3 billion prescriptions were filled in the United States. In that year, 2.2 million people were hospitalized because of ADRs, and of those, roughly 106,000 died. Fatal ADRs are the fourth leading cause of death in the United States, and these statistics reflect only the reported ADRs. Moreover, individuals taking several medicines daily, common among the elderly, are twice as likely to be harmed.[5] The Food and Drug Administration (FDA) Center for Drug Evaluation and Research current estimates are roughly the same as the 1998 numbers, indicating a lack of progress in remedying this serious public health problem. Notably, these estimates represent only a fraction of the actual costs, as they do not include unreported events. In other words, because each of us is unique, human biochemical variation is a formidable challenge to medicine's success. Variance in both genetic and environmental influences is an important factor in drug response. Scientific and technological advances promise to deliver personalized medicine, thus revolutionizing prescribing as we now know it.

Pharmacogenomics and Pharmacogenetics

Pharmacogenetics is the study of genetic variation and drug response. Its cornerstone is the ability to identify tiny genetic variations (single nucleotide polymorphisms or SNPs) that alter drug concentration during absorption, metabolism, clearance, excretion, and ultimately the clinical response. Scientists know, however, that the wide variability in people's response to drugs is *not* solely a function of their genotype. Emerging genetic subfields promise future tools to improve prescribing.[6] One such field, pharmacogenomics, examines the inheritable response to drugs over the entire human genome, considering the roles of gene products, such as RNA and proteins, and multiple genes in modifying a response. In investigating the molecular functionality of drug response, genomic technologies promise to improve drug development as well.

Clinical Applicability of Pharmacogenetics and Pharmacogenomics

Many people with common diseases don't respond to initial therapy. Determining the right medicine at the right dose can take several weeks, if not months. This lag time to therapeutic benefit can be costly. In the United States alone, data indicate that the majority of ADRs result from genetic variants influencing an individual's response.[7] While not all ADRs are the result of gene variance (ADRs can result from environmental influences, drug-drug interactions, and other biological processes), research indicates that roughly 50 percent were caused by gene variants.[8]

Ethical Concerns

The ability, however, to stratify individuals into treatment groups based on their likely response raises a host of ethical concerns. Among the numerous ethical questions raised are the following: What does fairness require for nonresponders, for example, people for whom available treatments are neither safe or effective? Will populations be left without treatment because they constitute too small a group for a pharmaceutical company to invest in treatment development with little if any return on investment, resulting in discrimination and related economic fall out?

Will physicians be obligated to offer testing and potentially be liable for adverse drug events if a test is not done? Will governments require testing to save health care dollars and in functionally coerce state-assistance enrollees to be tested? Will ethical issues exist within the doctor-patient relationship and outside it to broader worldwide societal concerns, such as who will have access to safer drugs, that is, only the rich? The following hypothetical scenarios are intended to highlight some of the most important ethical concerns.

Ethical Issue 1: Issues of Justice

Who should control access to new pharmacogenetic tests? Should this be the government, consumers, insurers, or physicians and other health professionals who prescribe medicines?

Important ethical questions involve determining *when* pharmacogenetic testing is good enough for public use, *who* ought to decide this issues, and *how* such a decision ought to be made. Tackling these issues involves lots of subquestions. Should government regulate pharmacogenetic testing to protect people? If so, should these be genetic diagnostics or identity tests? Should regulating agencies determine criteria for commercial availability or should individuals be free to buy and use tests even if they are not approved by the FDA? Should the individual maintain control to ensure genetic privacy from insurers and other third parties?

Deciding when pharmacogenetic testing is appropriate for commercial use undoubtedly will involve demonstrable proof of clinical validity and utility, but also will involve resolving competing interests of stakeholders such as balancing an individual's right to know against society's duty to protect its citizens. Patients, on the one hand, have a right to know information that can have a direct bearing on their health and may well demand access to testing despite cautionary objections from a conservative medical community insisting that testing wait for evidence-based studies. Still, on the other hand, patient demand likely will be mitigated by ability to pay, as health care insurers do not want to reimburse for tests until the clinical utility and cost-benefit ratios favor their pocketbook.

Practitioner acceptance will be influenced by (1) education in prescribing, (2) their degree of confidence in new approaches, (3) a need for approval from the upper echelons of professional medical organizations, and (4) the identification of a new standard of care. Pitting physician caution against patient demand requires broadening most discussions of harm and benefit. What physician would not want to prescribe drugs knowing that they will be far more beneficial than not, while avoiding liability risk associated with ADR? Such discussions primarily focus on the merits of scientific findings and the basis for proof, with less emphasis on possible adverse effects. This discussion is changing with increasing awareness of previously suppressed postmarketing surveillance data that indicate adverse events. Notably, we find ourselves at an interesting juncture at which medical care standards are increasingly evidence-based and yet the responsibility and cost of one's health is increasingly born by the individuals. We must address competing interests directly and properly frame issues for participatory discussion to ensure a just balance of those interests.

Ethical Issue 2: Issues of Privacy, Autonomy, and Informed Consent in Clinical Research

Within families, whose right to knowledge predominates? Should it be parental prerogative to always trump the consent of minors? How should we handle opposing rights of identical twins, for example, when one wants to no know their genotype and the other does not? What about family members who run the risk of unauthorized access to their private genetic information by test result implication from their relative's test results?

Genetic information is particularly vulnerable to privacy violations because it is informative of not only an individual, but also his or her relatives. The recently enforced Health Insurance Portability and Accountability Act (HIPAA)[9] of 1996 regulations could obviate some contentious privacy battles, by requiring all patients to expressly consent to disclosure of private health information and authorization of access to specific third parties, but HIPAA pertains only to the electronic transmission of health information. As noted earlier, access to genetic information via family history intakes that implicate blood relatives (without their consent) is a potent concern. The limit of an individual's right to privacy in this type of situation was itself tested a few years ago.

The ethical issues of informed consent and privacy are particularly complicated in pedigree research—that is, research at the stage of determining eligibility, which requires provisional pedigree construction, and at the stage when the study is actually under way. Concern focuses on the privacy of family members of the proband (research participant). Existing guidelines governing privacy protection were challenged in 2000 when the father of a potential research participant opened a letter addressed to his daughter, a consenting adult, and learned that her participation required disclosure of sensitive medical information about her father, who was not a consented subject in the study. The father objected to what apparently was nonconsensual disclosure of his medical information. He took his concerns to Congress, the National Institutes of Health (NIH), the Office for the Protection of Research Risks (OPRR), and other government agencies. Using the 45 Code of Federal Regulations 46 (45 CFR 46), "Federal Policy Guidelines for the Protection of Human Research Subjects," the OPRR determined that secondary subjects (like family members) fit the definition of human subject.[10] The OPRR determined that his written permission should have been obtained.

Among the difficult issues in this case is the fact that it challenges us to think about the weighted values we assign to first principles, namely, the right to privacy particularly in pedigree research, as well as in clinical family history taking. The question of who's right predominates is problematic because of the inherent potential for conflict. The protections and guidance that exist for research settings do not govern clinical settings, although similar issues can easily arise. Such issues are not amenable to applying a single quasi-mathematical formula and, not surprisingly, both institutional review boards and hospital ethics committees have come under fire in recent years for collective inconsistencies in decision processes and outcomes. The input from trained ethicists should continue to be sought if we are to achieve ethically responsible and defensible policy decision making.

Other ethical issues arise in clinical pharmacogenetic and pharmacogenomic research. Issues relate to the recruitment and selection of research participants and eligibility criteria. A core concern is whether it is ethical to select participants based on their genotype. While the argument can be made that genotyping constitutes one of many eligibility criteria, genotypic stratification of participants might lead to inequitable representation or possible loss of benefits in double-blind control studies. In addition, because genetic information is informative of relatives, the potential risks of gene-based discrimination and privacy violations extend to them as well as the research participant, raising questions about who is the subject? Or, how do we define risk? From whom is informed consent required, given that a research participant's genetic information is also informative of relatives?

Ethical Issue 3: Issues of Access to Testing, Privacy of Genetic Test Results, and Potential Discrimination in the Context of Commercial Use

Once individualized medicine is available, will it be used in an ethically appropriate manner?

Knowing whether a person may respond to a drug in ways that are safe and effective will enable patients to avoid medications that are dangerous or ineffective. Although desirable and cost-effective, if not also morally required and legally sanctioned, achieving widespread pharmacogenomic benefit, while minimizing harm, in practice may well be difficult. Consider the following hypothetical situation, which illustrates a wide range of ethical concerns arising in the clinical use of pharmacogenomics.

Hypothetical Case Scenario. Consider the following hypothetical, but plausible, case illustrating access, privacy, and potential discrimination concerns. Imagine that it is the year 2010. Patent M, a 42-year-old man of Scandinavian descent, presents to his physician with a general feeling of malaise. Five years before the visit, M was diagnosed with high serum cholesterol, which he attempted to control with a regimen of exercise and dietary regulation but without success. His physician then prescribed a drug therapy for him. After six months of therapy, there was only a modest lowering of cholesterol levels, so the medication was changed. After nine months on the second medication, there was still no marked effect. By the time M was able to see his physician again, a newer therapy had become available. This new drug had become the physician's favorite. The physician advised M to switch to this new drug, and the patient was eager to try it. Three weeks later, M returned to see the physician to complain of continued malaise.

Patient M may have been better served if he had undergone the following three genetic tests, the results of which could have provided valuable therapeutic management information:

- Test 1 determines whether the patient has a polymorphism associated with a predisposition for plaque development leading to coronary heart disease (CAD).

- Test 2 determines whether the patient has a polymorphism associated with a nonresponse to the new medication. A positive Test 2 indicates that the patient lacks an enzyme needed to metabolize the drug. The absence of the enzyme prevents the drug's absorption, resulting in its elimination without conferring any therapeutic benefit.

- Test 3 determines whether the patient has a polymorphism that indicates the presence of an enzyme responsible for metabolizing the dosage too slowly, thus making the drug linger in the body and creating toxicity to the patient, and causing possible skeletal muscle damage, which could lead to risk of developing renal failure caused by the excretion of myoglobin.

If M tests negative for Test 1, meaning that he does not have the polymorphism associated with plaque development, then his high cholesterol poses little or no health risk and medication to lower cholesterol levels is not indicated. If M tests positive, meaning that he does have the polymorphism, then he is predisposed to CAD by virtue of being a plaque maker and cholesterol-lowering medication is indicated as a preventive measure.

Patient M would like to resolve this issue once and for all. He is eager to be tested because his job status will be more secure if he can prove that he is not predisposed to CAD. Because his employer recently circulated health promotion pamphlets to all employees informing them of the benefits of pharmacogenomic testing, he now believes that he can avert what might be construed as a job risk if the result is positive. Learning which drug he should take would demonstrate to Patient M's employer (who is self-insured and so is his health insurer) that his compliance with a therapeutic regimen reduces his risk for CAD, thereby averting job risk.

The employer's motivation may be twofold: (1) to promote employee health to avoid low productivity caused by illness and sick days, and (2) to keep health benefit expenditures as low as possible. Ultimately, given that health care delivery is part of the free market, who will bear the burden of paying for the use of this new technology is far from clear.

The decision to undergo testing, while easy for Patient M, however, involves the wishes and rights of his identical twin, for whom M's results apply. M approaches his twin brother P to discuss his interest in being tested and hopes to reach a mutually agreeable decision. Brother P is close to M and believes that M has P's best interest in mind; however, P does not want to be tested. He just does not want to know. Complicating the issue is the fact that P, a recovering alcoholic, is receiving Medicaid, unemployed, and living in a halfway home. P's cholesterol levels happen to be nearly identical to M's. M believes the benefits of testing far outweigh the medical and societal risks. P disagrees and believes that he may be coerced into learning information about himself that he does not want and has no use for. He is afraid that M will decide to go ahead and be tested or that Medicaid will force him to do the testing to determine whether P needs medicine. Either way, Medicaid could save a bundle over P's lifetime if the information was known.

Further complicating the issue is that M's wife wants their three children tested, so she might know whether any changes in diet and lifestyle might improve their children's health and possibly reduce the need for drug intervention for her husband. Patient M, however, does not want their children tested.

Analysis. The case of patient M is designed to illustrate a significant number of competing ethical issues. Should a genetic test be required if the results can lead to better care and minimal risk of harm? Whose autonomy (or right to know) prevails, twin M or twin P? Medicaid's? And on what grounds? Assuming that a solid firewall can be put in place to maintain the confidentiality of M's test result, does that constitute adequate (or ethically defensible) privacy protection when real-world separation is problematic? Should parental prerogative for deciding what's best for their children extend to genetic testing? How should we balance the rights of individuals against other individuals (twin against twin, husband against wife, parent against child, employer versus employee, Medicaid versus enrollee?) regarding genetic testing? If doctor H is the primary physician for both M and P, to whom is his primary duty when considering the question of testing? Can the physician fulfill the Hippocratic Oath for both P and M? In the absence of criteria and guidelines, how can the doctor fulfill her/his obligations (which in this case conflict) in an ethically responsible manner? These dilemmas can arise in all genetic testing situations, though with one important difference: the risk-to-benefit ratio.

The risks of pharmacogenomics testing are in most cases minimal and the benefits enormous. The risk of psychological harm, for example, is far less substantial than from testing for a late onset monogenic disorder like Huntington disease, for which there is no effective treatment. It is possible that genes important in the pharmacology of a particular medicine may have pleiotropic effects, with specific results also conferring a predisposition for a different disease. Indeed, it's predicted that the standard of care will require pharmacogenomic testing and that physicians who prescribe without knowing a patient's pharmacogenomic makeup will be legally liable in the event of an adverse outcome.

Ethical Issue 4: Issues of Justice and the Just Allocation of Resources

In light of the cost shifting of health care, namely that health insurance rates continue to climb while coverage diminishes requiring individuals to pay more out of pocket for basic services, coupled with the long road to pay or receive reimbursement for new clinical genetic technologies, is it possible to ensure a fair distribution of the benefits and burdens of pharmacogenetics? Furthermore, can we ensure equitable global benefit sharing?

At issue is who will *really* receive the benefits of pharmacogenetic tests and personalized medicine down the road. Will the growing number of uninsured (now nearly 50 million) suffer more because they cannot pay out of pocket for tests and indicated treatments? Might state governments require Medicaid recipients to undergo pharmacogenetic testing to avoid spending large sums of money on drugs that don't work or on treatments to mitigate the adverse effects of drugs causing serious and even life-threatening events? This is not a far-fetched thought, as states in the past have implemented technology to solve certain social ails and currently operate with substantial budget deficits. For example, some states require the children of single mothers to undergo paternity testing to identify fathers ("dead-beat dads") to force them to pay child support. Clearly, ethical conflict arises here because mothers are essentially coerced into paternity testing (parental prerogative and voluntary informed consent do not then exist) and fathers are identified against their will. It's not hard to imagine the same situation arising in the future for Medicare recipients, who typically must take large cocktails of medications, before they can participate in the new Medicare drug benefits.

Similar issues arise for patients who have helped companies to develop new diagnostics and treatments by participating in clinical research. Often, the sickest patients, who most need new treatments, volunteer to participate in research to establish safety and efficacy levels required for market approval; however, once the therapies are commercially available, they are inaccessible because the exorbitant price is prohibitive or quickly drains lifetime medical insurance capitations. Were it not for the contributions from those individuals, companies would not have been able to develop or profit from new diagnostics or therapeutics; yet after having helped companies, those same people are locked out of the benefits of the new technologies. Access can equally be restricted by payers of health care, either third-party payers or Medicaid or Medicare. Such is the case for thousands of patients with Gaucher disease, who could benefit enormously from new treatments, such as Cerazyme, but they cannot afford the drugs and thus remain on less costly and far less effective therapies. Insurers happily pay for hip replacement in Gaucher patients but refuse to pay for the cost of enzyme replacement, which alleviates the disease progression that inevitably leads to the need for hip surgery, as well as other progressive disease symptoms requiring similar invasive procedures.

Realizing the disparity between their contributions to and rewards for drug development, individuals have sought just restitution. The first such case involved John Moore, who in 1976 was diagnosed with hairy cell leukemia and donated his tissue to research. His physician subsequently developed a cell line derived from Moore's tissues that was later commercialized by the University of California. In this particular case, the court found in favor of the University.[11] However, apart from the obvious financial inequities that create injustices, patients have recognized and sought other types of restitution for injustices arising in the process. Recently, for example, the Canavan Disease Support Group, which had been instrumental in identifying the gene, sued the drug maker to ensure access for future testing less they be priced out of it.

Ethical Issue 5: Pharmacogenomics on a Societal Level

Is population screening ethically appropriate?

Pharmacogenomic testing in many cases will satisfy the standard criteria for establishing a screening program, such as prevention of life-threatening reactions to certain drugs, assurance of a favorable cost-benefit ratio of testing and prescribing, and avoidance of particularly injurious ADRs, especially those who are costly to treat. Screening all individuals to determine whether the most commonly prescribed drugs are likely to be truly safe and effective would likely result in decreased health expenditures (by prescribing only those medicines that are truly safe and effective), decreased morbidity, and decreased mortality. Predicted benefit, however, does not ensure that screening is ethically acceptable, because many ethical factors involved in screening implementation would need to be resolved. For example, if access to the indicated alternative treatment was problematic, either because insurers refuse to cover the expense of therapy or individuals could not afford the best treatment, then instituting screening would be ethically problematic. If testing was a condition of insurance coverage, would an individual have the right to refuse? If test results must be shared with employers to ensure continuance of employer-based insurance coverage, would an employee have the right to privacy of medical information that might be construed correctly, or incorrectly, to affect the employee's future health?

Central ethical concerns are who gets to benefit from this new technology and who should pay for this technology, and they are a matter of fierce political debate. Different payer systems confer different answers to the question of who gets to benefit from pharmacogenomic tests and customized medicines. It has been suggested that under a single-payer system, all would benefit from this new technology, because the risks are spread over the community as a whole, but such is difficult to envision under our current free market or tiered systems. Given our current health care system, is it fair that access exists only for those who can pay for the tests and indicated treatments, regardless of insurance coverage? The likelihood is low that individuals would be refused health insurance because a particular therapeutic is unsafe or toxic to them. It is plausible, however, that Medicare or Medicaid would require testing as a condition of enrollment, because both programs have a vested interest in avoiding unnecessary health expenditures, such as the prescription medicines that don't work or the costs associated with treating adverse drug effects. This scenario raises numerous ethical concerns. Might these programs coerce individuals into genetic testing? If individuals refused testing, would they still be treated? If the standard of care required pharmacogenetic testing, what would be an ethically acceptable treatment? Would employers, many of whom are self-insured, discriminate against individuals who require expensive designer drugs rather than the cheaper versions (which work for many but not all people)? The cost of pharmaceuticals continues to rise and individuals are increasingly bearing the cost burden, and typically increased copays create unfavorable cost-benefit ratios for middle- and lower-income individuals. In other words, medical and economic incentives to make personalized medicine a reality do exist, but these forces alone will not ensure that benefits are accessible to all.

In fact, the socioeconomic context in which pharmacogenomics is developing has raised concerns about whether the field itself is ethically justified. Critics, primarily those who question the validity and utility of the entire genome enterprise (human, vertebrate, invertebrate, and plant), contend that 47 million people in the United States have hardly any access to basic, let-alone sophisticated medical care. This population is unlikely to have access to customized medicine if health care is delivered in a free market. It is hardly ethical, they argue, to develop medical technologies that could greatly benefit people if the majority is denied access. Though the cost of pharmacogenomics tests may be low, newly designed drugs are likely to be so costly that they are out of the reach of all but the wealthiest. They argue further that the manner in which the field is developing is unethical, because society is not deciding to develop this technology or how it ought to be used. Rather, private companies, which came into being often as a result of government-funded research and tax subsidies, are advancing the field despite what some argue are social forces demonstrating that benefits will not be

returned to taxpayers nor will taxpayers have significant decision authority with regards to use.

Conclusion

Despite our best efforts to anticipate and resolve ethical questions
stemming from the application of pharmacogenetic and pharmacogenomic technologies, additional unforeseen dilemmas will arise. The
issues discussed in this chapter are by no means the only ethical issues
associated with pharmacogenetics and broader adoption concerns. For
example, it is not difficult to imagine how pharmacogenomics testing
could generate far more personal genetic information than a mere drug
response result. High-speed easy microchip testing for effect A may
result in information about effect B as well. For example, the same
genetic variant associated with nonresponse to codeine in the future
might also be associated with late onset psychosis. The stakes involved
in using the technology, in other words, may be too high if, in seeking
to identify a probable drug response, one obtains additional information
having grave far-reaching implications, thereby complicating ethical
issues surrounding individual testing by orders of magnitude. Pandora's
Box is not yet open, but we are glimpsing what could be inside.

Notes

1. Examples include Herceptin, Iressa, and Gleevac.
2. de Leon, J., Armstrong, S., and Cozza, K., "Clinical Guidelines for Psy-
chiatrists for the Use of Pharmacogenetic Testing for CYP450 2D6 and CYP
450 2C19," *Psychosomatics* 47 (2006): 75–85; Gilbert, C. W., et al., "Targeted
Prodrug Treatment of HER-2-Positive Breast Tumor Cells Using Trastuzumab
and Paclitaxel Linked by A-Z-CINN Linker," *Journal of Experimental Therapeutics and Oncology* 3 (2003): 27–35.
3. The liver enzymes, commonly referred to as microsomal enzymes, are
now what we now call "drug metabolizing enzymes."
4. Isoniazid was biotransformed by acetylation and individuals were shown
to differ as much as sevenfold. Studies of Caucasian and Japanese twins demonstrated that individuals could be grouped into subpopulations of rapid and
slow acetylators.
5. National Institute of General Medical Sciences, National Institute of
Health, NIH Senior Health; http://nihseniorhealth.gov/takingmedicines.
6. Related fields such as toxicogenomics or envirogenomics study individual genetic variation in response to industrial chemical, occupational toxins,

pollutants, and other environmental exposures such as radiation, heat and cold, and food.

7. Grant, S. F., and Hakonarson, H., "Recent Developments in Pharmacogenomics: From Candidate Genes to Genome-Wide Association Studies," *Expert Review of Molecular Diagnostics* 7 (2007): 371–93.

8. Evans, W., and Johnson, J., "Pharmacogenomics: The Inherited Basis for Interindividual Differences in Drug Response," *Annual Review of Genomics and Human Genetics* 2 (2001): 9–39.

9. See www.hhs.gov/ocr/hipaa.

10. See Minutes from National Institute of Neurologic Disorders and Stroke Advisory Council meeting to discuss this issue.

11. *Moore vs. the Regents of the University of California*, 739 2d. 479 (CA 1990); http://www.richmond.edu/~wolf/moore.htm; http://www.ninds.nih.gov/find_people/nands/council_minutes_may2000.htm.

CHAPTER 6

Ethical Challenges in Biomarker Research and Biorepositories

The early 1960s saw the birth of large-scale DNA data banking when the Guthrie test became widely used and then mandated by states. Dr. Robert Guthrie, a microbiologist and pediatrician, and also the father of a child diagnosed with phenylketonuria (PKU), questioned whether the standard "diaper test"[1] diagnostic could be improved on, because it was not reliable until the child was several weeks old, at which point irreversible damage had already occurred. He succeeded in developing a presymptomatic diagnostic that detected blood phenyalanine levels by blotting blood from a small heel puncture of a newborn. The method proved not only accurate but also to be the best strategy for detection, namely, to test babies before they left the hospital. Thus, newborn screening was established. In 1963, the Massachusetts legislature mandated testing of every newborn. Because of its clear immediate benefit, soon thereafter, all 50 states and countries around the world required the screen.

At this, the inception of blood banking, the many ethical concerns that now challenge biospecimen collection and storage were not even considered, such as ownership, rights to further testing (such as rights to access a sample if not owned by another institution), or data-sharing issues. Moreover, in the early years of newborn screening, the consenting process was minimal, if at all. A newborn might be tested without the mother's knowledge, and consent did not include provisions for consenting to future research, disclosures to third parties, or waiving rights to generated inventions.

Today, ethical concerns surrounding genetic data banking have come to the foreground, because biorepositories of cell lines, DNA, bio-markers,[2] tissues, and the like are growing in volume and increasingly accessed by researchers around the globe. Each state, for example, owns a substantial repository of newborn blood, and this is but one of many types of biospecimen banks that exist globally. Conservative estimates indicate that there are more than 307 million tissue samples from more than 178 million people, with roughly 20 million new samples being added each year. Suffice it to say that today most, if not all, Americans have their blood or tissue banked somewhere, whether they have con-sented to it or not. With increasingly large amounts of data linked to specimens and the ability to globally link disparate databases, ethical challenges that lay dormant for decades are now beginning to bubble up with volcanic-like intensity.

Molecular medicine is not possible without substantial genetic research. To this end, there is an increasing need for large population studies to assess genotype-phenotype correlations, and hence large bio-repositories are being established all over the globe to further research. The Genographic Project, consisting of 10 research laboratories world-wide, was established to collect DNA from all populations, particularly those of the world's dwindling indigenous and traditional groups whose genetic identities and traditions remain relatively isolated, to map human migration patterns. The National Cancer Institute is establish-ing what it expects will be millions of tissue samples to map cancer genes. The National Institutes of Health is doing the same thing to identify human variation and track common disease genotypes and phenotypic correlations. Private companies are collecting samples to further disease and drug discovery, and disease support groups are aid-ing these efforts.

Genetic research offers unique opportunities, because DNA can be stored for future testing, unlike other types of biologic specimens (e.g., blood, urine, and skin tissue which have limited in vitro life spans). A collection of DNA samples represents an enormous amount of poten-tially significant information, which is why banking DNA has become vital for furthering genomics and related subfields. Unlike epidemiologi-cal research, the storage of DNA specimens for future testing long after an original consent was obtained raises unique ethical challenges. Fur-thermore, specimens can be informative of blood relatives, thereby extending the boundaries that define who is and who isn't a research subject. That is, if a biomarker is familial, the genetic information iden-tified applies not only to the proband but also to the other relatives who share that biomarker. In addition, the clinical significance of a biomarker is unknown at the outset and may remain so even upon the completion of the research, thereby, some argue, making valid prospective informed consent inherently impossible. Biomarker findings vary in their predictive value, with highly predictive information being more ethically charged than less predictive information, and it is this characteristic that further complicates consent. Moreover, the clinical meaning of a biomarker requires statistical validation and possibly even confirmation at a meta-analysis level, so disclosures in advance of statistical validation may be premature and lead to unnecessary harm. For these reasons, many argue that subject protections should be *more stringent* than those that apply to clinical research. For example, particular familial biomarkers that are highly predictive are likely to engender additional or heightened ethical concerns, if not only for familial grouping but for subgroups of the population defined by those biomarker findings. The storage of samples creates its own ethical concerns.

Historic Background

Guthrie spot collections, the first-ever DNA-based biorepositories, are now more than 40 years old. Mandated by all 50 states, all newborns are subjected to the spot test for anywhere from 9 to 50 inherited metabolic disorders in newborn screenings, depending on which tests the state conducts.[3] As of this writing only one state obtains consent for newborn collections.[4] Only here are parents legally able to exercise their rights and wishes about the ultimate disposal or retention of the spot for possible future testing.

From a public health view, molecular medicine is not possible without a solid understanding of genotype-phenotype correlations; in particular, how polygenic and multifactorial conditions develop and evolve. Guthrie spot repositories, therefore, represent one such collection of tremendous untapped value to researchers. Ethical issues surround the feasibility of testing these samples, though. Whether each state truly owns it collection and therefore can provide access is, arguably, unclear. Whether consent to further testing is required or could even be achieved constitutes another concern. Furthermore, were researchers to detect disease, they are under no obligation to contact the family with a diagnosis or set in motion indicated treatment, because therapeutic relationships, and their embedded obligations, do not exist between researchers and subjects. Research participants whose disease is detected in research are unlikely to get a diagnosis even if their specimens remain identified, and this represents a central ethical concern when treatments or preventive measures exist.

Notably, concern can extend to **anonymized** specimens as well, that is, samples no longer associated with any personal identifying information such that it is impossible to trace a specimen to its original donor. In the early 1990s, ethical concerns about identifying pathology in an individual's specimen but not informing them came to light when researchers found that women of Ashkenazi descent had a significantly greater chance of inheriting a BRCA1-2 mutation than women in the general population. Individual study participants had no way of knowing whether they harbored mutations that increased their risk, which arguably could have prompted them to seek preventive measures.[5] Findings are commonly reported in aggregate, which protects an individual's privacy but offers no direct clinical benefit. Moreover, when the researcher is also the participant's physician, the distinction between research subject and patient would seem blurred. In research, physicians do not have a therapeutic relationship with their patients, and individuals lack the rights they have in a doctor-patient relationship.

As of this writing, a person's rights over their DNA (arguably among one's most essential characteristic) are less defined than one's right to property (arguably less essential), such as one's home, car, or finances. The law affords people the ability to protect themselves, including their blood, tissue, and DNA, from harm (harms that are material or psychological), but only as long as it exists within in one's body, a precedent set by the landmark Moore case.[6] The case addressed the question of who has rights to an individual's blood, tissue, DNA, and the like once it is removed from the individual's body. John Moore's spleen was removed in 1976 as part of his hairy-cell leukemia treatment. Moore's doctor, David Golde, a cancer researcher at UCLA, identified several valuable proteins and notably a rare virus, which was thought possibly capable of leading to HIV treatments, in Moore's cells. Without Moore's knowledge or prior consent, Golde turned Moore's cells into a cell line, or self-perpetuating clones of the original cell, and patented it. In 1983, when Moore was asked to grant the University of California all rights to any cell line, or potential product developed from his blood or bone marrow donations, he became suspicious and subsequently sued. Among the ethical problems arising in the case was the fact that Golde had two disparate relationships with Moore—that of his physician who made his diagnosis and treated him and that of cancer researcher—each with different, albeit conflicting, obligations.

The issue of ownership, and hence rights to one's bodily parts, so far has turned on a definition of property rights, such that one's right to ownership ceases once the specimen is removed from the body, regardless of whether removal was consensual. After several court cases and appeals, the California Supreme Court, in 1990, ruled against Moore stating just that. Furthermore, the court ruled that Moore didn't and couldn't own his cells because ownership would conflict with the patent. Given that Golde had transformed Moore's cells into a cell line, they were no longer Moore's cell but rather Golde's invention. This has remained the definitive statement on the issue. That is, once property is removed (disembodied), the Common Rule (the federal law governing consent to human research) no longer applies. Moore did prevail, however, on two of eleven counts in the case: lack of informed consent and breach of fiduciary duty. The court ruled that Golde should have disclosed his financial interest to Moore as well as the lack of existing regulation concerning consent and ownership. However, the court worried that giving property rights to patients would "hinder research by restricting access to necessary raw materials" and in effect lead to requiring researchers to purchase each and every cell, or part thereof.

Although it is illegal for individuals to sell human body parts such as organs or tissues (removed from the body)[7] for transplantation or medical treatment, it is legal to sell DNA, biomarkers, and tissues, which raises its own ethical issues. For the most part, it's someone else who's selling your specimens. Once your tissues leave your doctor's office and are sent to the lab, they are no longer yours. The lab has the right to extract what it deems appropriate for either diagnostic or research purposes (say a particular 5 centimeters of specimen) and throw away all the rest or store it for future use. Technically, anyone can rifle through discarded samples and sell them for whatever purposes, though biohazard waste is off limits. In fact, some argue that it is morally imperative to keep this practice in place to ensure that research continues to achieve a greater good, such as targeted prevention strategies and therapeutics. David Korn, a senior vice president of the Association of American Medical Colleges, has been quoted as saying, "I think people are morally obligated to allow their bits and pieces to be used to advance knowledge to help others. Since everybody benefits, everybody can accept the small risks of having their tissue scraps used in research."[8] Whether "everybody" truly benefits is debatable, particularly in light of the increasingly large number of the uninsured. Nonetheless, in his view, consent requirements are secondary to a more primary public responsibility, which is to advance medical science for the well being of all.

The goal of obtaining new knowledge is itself a normative proposition. Advancing medical genetics for its own sake, independent of its utility, may be socially desirable, although the primary value is likely to lie in its utility. Independent of ethical concerns about the application of new genetic knowledge are ethical concerns about the methods used to achieve advances. Nazi-era human research, as a case in point, violated basic codes of ethical treatment of subjects, and thus led to the establishment of human subject protections (the Nuremberg Code), including enforceable regulations designed to ensure the ethical oversight of clinical research. Requiring that the methods used to obtain new medical knowledge meet ethical standards shows that research processes are as important as the value of new knowledge.[9]

Standard Ethics Concerns in Biomarker Research

The standard ethical concerns involving clinical research using human subjects apply to clinical genetics research. Voluntary informed consent, confidentiality, privacy, and security of a sample and its information must exist and comply with state law. Ideally these conditions also meet recommendations set forth by the American Association of Medical Colleges' Institutional Review Board (IRB) certification program[10] as well as specific national and international bioethics organizations. Duties to participants must be met (whether these are investigator or institutional obligations) and results must be communicated to sources and, in some cases, to family members as well (see Tables 6.1, 6.2, and 6.3).

Institutional Review Board Oversight

IRBs and Institutional Ethics Committees provide oversight and exist to help ethical quandaries before and after they arise. Such boards take advisement from the Office for the Human Research Protections (OHRP), within the Department of Health and Human Services (DHHS). They are subject to random audits to ensure compliance with federal guidelines put forth by the OHRP. For the most part, however, they are not governed by state laws or regulations. These boards have come under increased scrutiny of late for instances of negligence or ethically substandard oversight.[11] Recent audits have found an array of offenses, including embarking on the research without prior IRB approval in the mistaken belief that approval was not needed for

Table 6.1
Standard Ethical Concerns

Consent	Confidentiality	Harm	Duties
Recruiting	Requirements	Quantity vs. quality	Conflict
During study	Norms/ethical practices	Type vs. degree	
Communication		Known vs. unknown	
Interpretation			
After study			
Future study			

anonymized data. The IRB, in one case, flagged the researcher but failed to report the incident to the federal health officials. In another case involving clinical experiments in China, the IRB asked the researchers to provide a plan for handling potentially suicidal subjects, pregnant women, and those stigmatized by their families and communities, but permitted the study to move forward before receiving and reviewing answers to these questions. Two years ago, the OHRP criticized three Harvard-affiliated institutions for lax patient protections during a genetic research experiment in China. This offense was considered by OHRP to be far more serious that those mentioned above, and

Table 6.2
The Human Genome Organization's 1998 Recommendations for Sample Storage

Consent
    Should specify uses of sample and its information
    Indicate whether sample and its info will be identifiable, coded,
      or anonymized
    Caution about irreversibility of anonymization (e.g., there will be no way
      that information can be shared with subjects other than as aggregate data)
Confidentiality
    Procedures to ensure confidentially must exist
    If high risk, family members should have access
    Sample may be destroyed at source's request if no need for relatives' access
    Except where law requires, no third party disclosure is permitted
    No disclosure without consent
    International standardization is necessary

Source: http://www. Gene.ucl.ac.uk/hugo/sampling.html.

Table 6.3
National Bioethics Commission Recommendations

A. Unlinked samples may be exempt from IRB:[a]
   i. if the process used to unlink is effective
   ii. if unlinking will not lessen the value of the research
B. Coded and identified samples:
   i. data banks and researchers share responsibility for ensuring that research is designed and executed in a manner that protects subjects from "unwarranted" harm
   ii. IRBs should require investigators to:
      1. provide a thorough explanation of design and execution in a manner that minimizes subjects' risk
      2. explain how medical records will be accessed
      3. provide a full description of how procedures to maximize privacy protection and minimize inadvertent disclosure
   iii. disclosure should occur only if results are scientifically valid, if results are significant to the source's health, or if a way to ameliorate health concerns is readily available[b]
   iv. if findings are disclosed, appropriate medical advice or referrals should be provided
   v. regarding publication of requests: journals should adopt a policy requiring all human subjects research results publications to specify whether research was conducted in compliance with the Common Rule, and this should apply to all research whether privately or publicly funded or otherwise exempt
C. Consent:
   i. form should explain collection, storage, and use procedures, including possible future use, in detail—for example:
      1. the possibility of future research participation
      2. the possibility of permitting only un-identified or un-linked use of biological materials
      3. possible use in one of several studies and no future contact
   ii. research and clinical consent should be obtained separately
   iii. waivers for minimal risk or impracticality
D. Communication of results:
   i. IRBs should develop guidelines for results disclosure
   ii. investigators should address all issues in research plans
   iii. disclosure should occur only if results are scientifically valid, significant to the source's health, or if a way to ameliorate health concerns is readily available
   iv. if findings are disclosed, appropriate medical advice or referrals should be provided

(Continued)

Table 6.3 (*Continued*)

v. regarding publication of requests: journals should adopt a policy requiring that all human subjects research results publications to specify whether research was conducted in compliance the Common Rule and this should apply to all research whether privately or publicly funded or otherwise exempt

E. Potential harm to others:

   i. researchers should anticipate possible harms and consult groups about study design

  ii. if research on unlinked samples poses significant potential for group harm, exemption should not be granted

 iii. if research poses specific risks to groups, risk should be disclosed

---

[a]The commission does not recommend who ought to determine whether the process to unlink is effective.

[b]As a practical matter, a time limit after which disclosure is not reasonable is not specified, thus contributing to the ad hoc nature of implementation.

Source: http://www.bioethics.gov.

notably was brought to the OHRP's attention by a whistle blower. These types of issues are likely to be found in any institution conducting clinical research.[12]

Informed Consent

Informed consent is now a legislated practice in every state. Not only is it immoral for researchers to fail to obtain informed consent, it is *illegal*. Failure to obtain *proper* consent constitutes battery. Proper consent entails researchers providing to potential human subjects (before participation) all known potential risks and benefits of participating in that particular study, including physical, psychological, and when obtaining hereditable information, potential societal risks such as possible unauthorized access. Furthermore, for the consenting process to be valid, potential subjects must comprehend the full nature of these conditions and their implications. Typically, researchers are encouraged to elicit feedback and questions from potential participants to determine whether participants truly understand the risks and benefits involved.

The consent process, while designed to maximize understanding of both risks and benefits to ensure ethical decision making about research participation, is limited by an individual's ability to comprehend risk. Individuals vary greatly not only in their perceptions of what is and is not risky, but also in how they react when faced with the risk of adversity. The consenting process lacks built-in mechanisms to protect against misunderstandings or misperceptions. Aggravating this problem is the public's relatively low level of science literacy,[13] particularly in understanding odds ratios. Guidelines for consent forms exist and IRBs approve forms before their use, but there are no objective criteria for determining the adequacy of consent forms, nor the consenting process. Some studies have multiple informed consents, one for each aspect of the research, including but not limited to follow-up contact with or without results. Although IRBs approve consent forms, an IRB approval, like that of a counsel opinion, is just a reasoned opinion that may prove substantially different from the reasoned opinions of other IRBs, which evidence demonstrates can differ vastly.[14]

Clinical genetic research, has, for some, raised concern that the paradigm for individual consent is no longer adequate, because research can require family history and results can implicate blood relatives or define subgroups of the population, who presumably have a right to determine whether their information should be available. Whether the consent obtained is truly voluntary and informed can be questionable at any stage of research or clinical application. Particularly when studies offer large sums of money for patient participation, payment is clearly an inducement to participate. Because financial inducements carry more weight with poor people, they tend to participate more than rich people, and thus generate selection bias. This imbalance might suggest that the poor may be less concerned about risk burden than they are about financial gain. They need the money. Payment as an incentive, then, can be viewed as compromising the voluntariness of consent.

These concerns can pervade the recruitment and research processes so that, at any point, a participant could announce that they no longer understand or voluntarily agree to continue, claiming that what is happening is not what they consented to at the outset of the study. As a practical matter, the right of refusal, or the ability to withdraw from studies, without incurring any ill effects is standard in consent. However, when specimens are deidentified or anonymized, or exist in banks in foreign countries, ensuring their withdrawal is complicated and potentially problematic. While poststudy contact that is not explicitly discussed in a consent form is generally forbidden, except in some rare circumstances in which failure to recontact is likely to impose great harm to the participant, researchers do occasionally attempt to recontact subjects. In cases in which subjects might greatly benefit from new knowledge, researchers may be obligated for moral reasons to do so. In practice, however, consent to possible poststudy recontact is not always provided in the original consent process. Furthermore, recontact may compromise an individual's privacy in ways not authorized by the original consent process.

This problem is particularly acute in the case of biomarker research, in that such disease markers may come to be understood as being pleiotropic, which could mean that the marker is associated with multiple effects, some beneficial and others harmful. Specific ApoE alleles, for example, indicate a risk for both Alzheimer disease and cardiovascular disease. In the latter case, knowing one's alleles is arguably beneficial, because preventive measures can be undertaken to lower one's risk. In the former case, such knowledge is arguably harmful, in that it creates significant psychological burdens in the absence of preventive or therapeutic measures, and such knowledge may not have been originally consented.

Of particular relevance to biomarker research is whether the consent is generic or specific to the hypothesis to be tested. This is a significant concern because ancillary research is often conducted concomitant with the original research, and the original consent can only be as thorough as the original aims of the study. Among other things, this means that if changes occur during the research, which are not automatically sanctioned by a prior consent (in most cases the prior consent did not provide blanket approval to any or all research changes), an ethical question arises about whether researchers are obligated to re-consent all participants. Because even the best and most specific consent process may become inapplicable for future testing with new discoveries, the limits of consent are always in question. Many debate the importance of participants maintaining control over determining what research is done on their stored samples; others arguing that subjects ought to be able to consent to certain research areas and dissent from research on other areas.

While no overarching recommendations or guidelines pertain, other than the need to include *possible* future recontact in the original consent form, uncertainty exists as to when data indicate the need for recontact. Potential participants may be difficult to contact if follow-up has not been maintained and contact links have been broken because of changes in residences, death, or some other event. An inability to recontact may introduce bias into the study. Studies that involve multiple genes or other related factors could theoretically require consent for every research question, which would generate an almost continuous stream of consent requests to the participant, creating an undue or questionably appropriate burden to the individual, as well as causing an administrative hassle. Other issues concern whether there are or ought to be limits to reaching back to subjects, particularly if they consented to subsequent recontact in the original consent. Should a conventional seven-year statute of limitations apply, or should a time limit be lower or greater, especially when considering the potential for enormous benefit resulting from research? Ought any limit to be imposed? In other words, in the absence of recommended time limits, it would seem unethical for limits to be set on an ad hoc basis. Importantly, if the data are proprietary, as in the case of postmarketing data, sources often are not fully aware that the information they could be asked to donate is proprietary and unlikely to be of direct benefit to them. Yet, failure to obtain new informed consents may expose the researcher to allegations of ethical impropriety. Furthermore, because biomarker information might be clinically relevant to the participant, there are ethical issues involving whether the investigator should disclose results to the participant (because they may benefit the participant or lead to an improved health outcome), even if the participant was not counseled before the study about the possible relevance of results. While research institutions have access to qualified bioethicists and several hospitals have internal ethics committees to handle internal quandaries, it may be the case that even these professionals are not adequately trained in the relevant genetic or ethical issues to be able to offer competent advice.

Consent in Nonnative Languages

The potential for less than fully autonomous decision making particularly exists in trials involving nonnative English speakers, who may not fully understand the protocol or who may abide by cultural norms that mitigate U.S. norms of autonomous decision making. When potential participants are nonnative English speakers, interpreters are typically needed. Involving interpreters in the consent process generates ethical concern about whether translations will convey directive or coercive communications. Interpreters need not be trained (nor certified for competence) in genetics, ethical issues in genetics, genetic counseling (which could help them avoid using coercive and directive language), or the subtle meaning differences in choosing certain words with particular nuances, each with their own ethical implications. Ensuring that such a potential participant truly understands or decides voluntarily can be problematic because stop-gap measures to ensure that the information is presented as neutrally as possible do not exist. When cultural norms governing individual decision making clash with the American sense of autonomy, obtaining ethically appropriate decision making can be difficult to assess. For example, in some cultures, it is morally acceptable for men to decide for the women or for younger relatives to decide for an aged relative, and in other cultures, it is ethically appropriate for older relatives to make all of the family decisions and therefore decide on behalf of the subject. Conflicting norms governing what constitutes ethical decision making make it difficult to determine whether conditions of informed consent are satisfied. Most important, these concerns are particularly problematic when children who are fluent in both English and the family's native language are used as interpreters. Though convenient for parents, this practice raises concerns about violating the parent's privacy (even with consent) as well as the appropriateness of involving minors in intimate personal details or other highly sensitive adult medical information.

Potential for Abusing Consent Privileges

Ensuring that consent is fully informed and voluntary does not end concern about ethical impropriety, because the potential to abuse consent privileges always exists. Arguably, the potential for abuse is greatest when researchers and consentees do not share the same native tongue. A case in point occurred in 1990. John Martin, an anthropologist at Arizona State University, approached the Native American Havasupai Tribe in Arizona about donating blood specimens to research a diabetes epidemic on their reservation. As initially presented to the tribe, the project was designed to educate tribal members about health eating habits and obtain blood samples to screen for diabetes and locate genetic variants, similar to research done among the Pima populations of southern Arizona in the early 1980s. Shortly after Martin's initial contact with the tribe, the university established the Medical Genetics Project of Supai. Over a decade later, in February of 2003, a tribe member questioned whether specimens were being examined for purposes other than those originally agreed to by tribe members. An attorney investigated this question and concluded that beginning in 1991 only oral consent statements were exchanged between researchers and donors. In other words, consent forms were not presented and written consent was not obtained. In 2004, 52 members of the tribe filed a $25 million law suit alleging that blood samples were collected without obtaining informed consent, without Arizona State University IRB approval, and in violation of federal and state laws and regulations. Furthermore, the suit claims that the specimens were used for far more than the diabetes research initially agreed upon, specifically schizophrenia, in-breeding, and migration patterns.

Consent Is Limited by the Location of "Your" Specimen

Consent is limited by the location of your tissue. As evidenced by the Moore case, while still in you, you have leverage in terms of consent and future use. Once removed, your rights and consent requirements cease. The ethical rub is understanding what it means to be fully informed so that your consent is in fact valid. For the most part, you can only exercise such leverage if you know the value of your tissue or cells. As nonexperts, most of us have no idea whether or not our tissues contain any uniquely valuable features. In fact, the paradox is that one cannot know in advance of doing the research. And yet the consenting process does not require researchers to disclose to subjects whether or not their tissues are potentially of great value.

Ted Slavin,[15] a hemophiliac in the 1950s, had been exposed repeatedly to the hepatitis B virus as a result of his clotting infusion treatments. He was wholly unaware of such until the 1970s, when his physician informed him that his blood contained a high concentration of hepatitis B antibodies. Slavin realized that his cells were worth a lot of money, as researchers were trying to develop a hepatitis B diagnostic test but required a steady supply of antibodies to do so. He began selling his serum to anyone who wanted it. However, much as Slavin wanted the money (he actually needed the money because his hemophilia has resulted in disability and many lost jobs), he wanted someone to cure hepatitis B more than satisfying his own greed. He identified the researcher most likely to succeed and proposed donating his antibodies for free. The difference between Ted Slavin and John Moore is that Slavin owned his tissues, he sold his blood while it was still in his body. Moore didn't know the value of his tissue until it was already removed from his body. An important difference between their situations hinges on informed consent. Slavin was told that his tissues were valuable and that someone might want them. Moore was not. The Slavin case is interesting for several reasons among which is that despite the financial value of his cells, Slavin displayed an altruistic drive to advance medicine to improve the lives of others. His relationship with the researcher also illustrates a reciprocal trust and fiduciary obligation that does not often exist between participants and researchers. Arguably, exercising consent in the former situation offers greater opportunity for realizing fully informed consent than in the latter situation, in which subjects hardly know, and have virtually no relationship, with the researchers working with their samples.

Who Controls the Consent: The Researcher or the Institution?

The case of Dr. William Catalona, which is pending as of this writing, illustrates this problem. (Most people believe that the institution is the proper recipient of consent, not Dr. Catalona, because for the most part institutions own employee work, which is the basis for compensation.) Dr. Catalona, a world-renown prostrate surgeon, began collecting prostate tissue specimens in the late 1980s while at Washington University in St. Louis. Committed to fully informing his patients, his consenting process included not only consent forms but alerted patients to quarterly newsletters that provided updated study findings. "His" collection of prostate cancer samples is one of the world's largest. Quotations surround "his" because a St. Louis court is currently sitting on the question of who owns the collection, Catalona or his former employer Washington University.[16] The University's suit against him originates in a dispute about whether Catalona was entitled to give free tissue samples to a biotech company, wherein the University would receive no compensation of any kind, no fees for sample usage, no claims on inventions, and so on. Of note, though not crucial to the point, is the fact Catalona has not patented any aspect of his research. The only potential benefit to him was an authored publication. Washington University argues that it invested millions of dollars in developing and sustaining Catalona's collection and that it not only paid his salary, and his health, liability, and malpractice insurance, but also owns any of his intellectual property, as per his employment contract. Washington University, in other words, claims ownership to the biorepository. When Catalona moved his lab to a different institution, he sent letters to 10,000 patients requesting their permission to release their samples to him (moving them from Washington University to Northwestern University, the new institution). The majority of his patients authorized the release of samples to Northwestern. Washington University is denying their requests on grounds that the patients do not own the samples and have no rights over them.

The question of who owns the samples, the patients, Catalona, or Washington University remains, at the time of this writing, unresolved. Part of what makes this case of particular interest is that Catalona's patients believe that they've entrusted their samples (in the words of one patient, his DNA and thereby that of his kids and grandkids) *only* to Catalona, who, incidentally, has displayed a noteworthy lack of interest in financial gain. Patients, who entrusted only Catalona with the authority to use samples as he saw fit, have voiced concern that Washington University's claim to owning their body parts, and thus a financial interest in those samples, makes them feel demeaned and comodified, as if the University has substantial control over their lives and all their future lineage. Unbeknownst to the patients, Washington University distributed samples to other scientists for research, which clearly the donors didn't know about or consent to. The University's action raises a particularly vital ethical issue, namely, that a participants decision to withdrawn from research is not the same as a directive to anonymize the sample and its information, thereby enabling research to continue on the sample as thoroughly cut off from its origin.

The issue of consent gets quite complicated in a hurry. While it would seem ethically necessary to give subjects as much autonomy as possible and to make consent forms as specific as possible, in practice this is not feasible. How can researchers function in a system in which each and every one of billions of specimens has unique specifications? For example, specimen A can be used to commercialize hereditable forms of breast cancer but not lung cancer induced by smoking and other bad health habits. The magnitude of the problem is illustrated by what one consents to, which may come to have negligible applicability based on future understandings of how mutations play out in heterogeneous tissues.

Confidentiality and the Security of a Sample and Its Information

Consent issues lead to concerns about confidentiality. Participants are assured (typically in the consent process) that their personal information is private, confidential, and inaccessible to anyone else. This is true whether the participant's sample is identified, deidentified, or anonymized. Whether they are stored for future testing or destroyed after testing is completed, samples are either identified, deidentified, or anonymized. *Identified samples* contain personal identifying information, such as a name, medical record number, social security number, or the like. *Deidentified samples* have been stripped of personal identifiers, such as name, birth date, address, social security number, and so on, and are given a code number for identification purposes. Anyone handling the sample or its information knows the code name but not the individual's name. The linkage, or which codes refer to whose samples, is kept strictly confidential and private according to various regulations and laws, and this code is functionally inaccessible to anyone, except possibly the person in charge of maintaining the linking system. *Anonymized samples* have not only been deidentified, but also lack any tracing mechanism enabling someone to link a particular sample to its owner. In other words, tracing a sample to its owner is impossible with fully anonymized samples. With increased data sharing and global linking of databases, concern now exists that software could undo some of the privacy protections afforded by anonymization.

The Health Insurance Portability and Accountability Act (HIPAA, the federal law protecting individual privacy of health information—genetic information is one of many types of private patient information—that went into effect in 2004) specifies certain privacy requirements in transmitting electronic communications, and HIPAA is not unrestricted in its scope or applicability. For the most part, HIPAA is strict enough to prohibit third parties from access to identifying information, even through the Freedom of Information Act (FOIA), which permits third parties to access identifying information in federally funded research. Notably, FOIA enabled tobacco companies to identify individuals who had participated in studies demonstrating the danger of smoking.

Although institutional procedures are usually put into place to limit unauthorized disclosures, ensuring strict confidentiality is of heightened importance in the context of multinational or multisite study information exchange. The likelihood that multiple IRBs have exactly the same understanding of this issue is low and, hence, the informed consents may not be equal throughout the study population.

Privacy

Protecting the privacy of biomarker data is increasingly difficult because of competing requests for specimen use. Many different investigators want specimens for many different studies and different commercial applications. The growing demand for specimens, coupled with the increasing ability to link disparate databases globally, suggests that privacy protections may be harder to secure in the face of mounting complexities.

While this system would seem adequate to both investigators and participants, privacy is not necessarily protected by it. Anonymity and privacy are not identical. Anonymous samples may not reveal sources' identities, but that is not the same as keeping the information private. A sample's meaning and clinical relevance (i.e., its epidemiological utility) are valuable only if shared or applied to other data or findings. Herein lies the rub. Sharing information, even when all identifiers are removed from the information, can be viewed as a violation of strict privacy. The problem is perhaps best illustrated by the recent Northwestern Memorial Hospital/Department of Justice (DOJ) case. The DOJ, as part of a proceeding against stem cell research, tried to obtain deidentified abortion data from Northwestern Hospital.[17] The hospital data that the DOJ wanted was anonymized ("stripped of identifiers," such as name, age, and other personal and demographic information so that ownership was virtually untraceable) and was requested as a way to identify potential stem cells. However, identity removal and disclosure of results in the aggregate only did not satisfy the patients' need for privacy. Even this form of presenting the data, the patients argued, amounted to a privacy intrusion. Arguably, the controversial nature of abortion data contributed to the perceived sensitivity of the medical data. This point is relevant to biomarker research. The controversial nature of certain research areas, such a psychiatric disease, aberrant (addiction), violence, or intelligence is likely to raise the bar on what is considered intrusive and what is not. Privacy in controversial contexts may well have to carry greater weight than similar issues in less controversial contexts.

Ethical Issues Related to the Specific Biomarker and Its Use

Beyond ethical concerns pertaining to any biomarker research are those that pertain to specific biomarkers and their specific uses. Why use this biomarker? Is this marker the best strategy or are other strategies available? Are there any unique ethical issues that pertain to this particular biomarker? Is the use of a particular biomarker ethically defensible? Could the expected information could be gleaned from alternative, more ethically acceptable methods? Is the proposed research needed and on what grounds is it justified? Is this research the best use of scarce resources? Are the scientific methods proposed appropriate for the research question? Could some if not all of these scientific or public health questions be answered by using different methods?

Ethical issues also arise from failures to plan or budget for unanticipated events. Such issues can arise in any clinical research, but their occurrence in biomarker research is particularly ethically troublesome because the stakes involved may be higher. Unanticipated events include situations like (1) an unexpected need to retest one or more subjects, (2) a need to provide counseling (including but not limited to genetic counseling) as a result of unexpected effects on the source (proband or index case) or groups related to the source, or (3) the need to obtain a diagnostic evaluation. These and similar occurrences are rarely, if ever, budgeted for, but they must be cared for properly if they arise.

Research Design

Ethical concerns can involve the specific design of the research. Typically, a study design is cross-sectional, prospective, or case control. *Cross-sectional studies* involve healthy people and typically are completed in a short period of time. Results most often provide insight into the potential risk of an exposed group as a whole and sometimes include individual risk. Notifying people of their results of these studies is common. In *prospective studies*, everyone is healthy and it's understood that results will not be available for a relatively long time. *Cohort studies* of identified high-risk individuals are more likely to inform the subjects of results. In *case control studies*, half of the subjects are already sick and the other half are randomly selected controls without an a priori (i.e., known or definable) risk. These subjects are generally not notified of their results. From an ethical standpoint, how can we justify such discrepant practices with varying implications to index cases and groups related to them? It is not clear that it would be ethical even with an informed consent, which doesn't offer disclosure of results, to withhold information that could be valuable in terms of the health of a subject. This speaks to the conflict that the physicians and researchers have with subjects and patients.

Disclosure and Communicating Results

Biomarker research raises another significant question—that is, whether, and if so when and why, investigators have a duty to a subject's *family* (or member at increased risk) if the biomarker is hereditable. Frequently, such possibilities are not contained in consent forms, and in the absence of recommendations, they are likely to be decided on an ad hoc basis. Breaking participant confidentiality to involve a participant's primary care physician is ethically suspect but may be ethically appropriate if a greater good (i.e., improved health or lowered risk of disease) is achieved. Furthermore, the issues about who the purveyor of results ought to be—the funder, the investigator, or participant's physician—is crucial, particularly when biomarkers are highly predictive of a serious risk of disease.

Consider the following: In 1983, researchers investigated whether blood lymphocyte proliferation in response to bertrandite ore exposure was a marker for predisposition to chronic beryllium disease, with the caveat that the disease is not possible in the absence of exposure.[18] The study assessed beryllium processors who were handling bertrandite ore at a mining operation in Utah. The company paid the lab directly for test results. A total of 16 percent of the workers tested were positive for the biomarker. Because they lacked clinical symptomatology consistent with disease, the workers were told nothing. The company refused to tell the workers their test results for two reasons: (1) the testing was experimental, and (2) workers lacked abnormalities indicative of disease.

Here a conflict existed between the employer's right to own and control information it paid for and the workers' right to know that information. Furthermore, a cardinal rule of disclosure to participants was violated. Researchers wanted to disclose results to the workers, but the company initially refused them access to the plant. The union intervened and eventually results were disclosed, although over the company's objections. Researchers also claimed that another cardinal rule of clinical medicine was violated, namely, to repeat testing to ensure accuracy, which in this case wasn't done. Subsequent to these actions, the company voluntarily revised its procedures to reduce worker exposure to the ore, and repeat tests revealed greater accuracy, including that some previously positive individuals were now negative. Since this incident, cases against employers whose workers were exposed to beryllium have increased considerably, with well over 70 cases in various state and federal courts.[19] A U.S. registry of cases as well as a Beryllium Help Center now exists.[20]

A separate concern applies to the process of communicating study results regardless of whether disclosure is to the proband or the proband's family member. In general, results ought to be communicated as objectively as possible and disclosed in ways that ensure that the proband and their family members know that these biological results do not necessarily carry more weight than other types of information obtained in the course of research (i.e., information from questionnaires, monitoring procedures, or other methods used). Doing so guards against a knee-jerk reification of genetic information, which in the case of biomarker research is not meaningful except in the context of other biological and environmental factors. Despite the plethora of genetics education available to both practitioners and consumers, the need to defend against genetic determinism remains—and is likely to remain for some time—a real need. Moreover, ensuring that integrity of the communication process for non-English-speaking participants can be highly problematic. Finally, because of the heterogeneity between studies/clinical settings and the cultural norms of participants, it is essentially impossible to establish uniform rules defining ethical practice in every application of biomarker research.

From Research to Clinical Use

Finally, ethical concerns are likely to arise in the transition from research to clinical use. Clinical significance will nearly always be a concern for biomarker research unless or until such significance is demonstrated by rigorous trials or usage becomes established practice. Among the issues likely to arise are the following: How strong is this information versus other clinical markers? How ought clinicians handle familial markers? Is there a duty to warn relatives? What does it mean to disclose in a "timely manner?" Arguably, it is unethical to communicate after the time has passed to do something that would have made a clinical difference. Of vital concern is whether advice should be given in the absence of a standard of care. An ancillary concern is who ought to be the purveyor of advice, if advice is in fact warranted and ethically appropriate?

Community Rights versus the Rights of Individual Members

Another ethical dilemma could arise in a situation in which the state could benefit from the genetic test results of a sizeable population. For example, if a biomarker for a cancerous condition resulting from exposure to an asbestos-like substance was discovered in a majority of state buildings, the state, which is likely to be self-insured, may want to test all employees to remove those who test positive for the biomarker, to reduce its potential health care costs and lost productivity as a result of missed work. The state, a self-insured employer, would be exempt from the Employment Retirement Income Security Act (ERISA)[21] and other federal laws designed to protect workers from discrimination or similar adverse events, based on a test result that is known to the employer and the insurer (which in this case are one and the same entity).

As purveyors of Medicaid, a similar situation could involve the state, which might want all enrollees tested as a condition of receiving treatments for health-related reactions to exposure, such as chronic beryllium susceptibility testing. Medicaid, like the state employer, could require testing as a condition of continued enrollment or employment, with failure to comply grounds for benefit exclusion. Furthermore, Medicaid could require (enforceability issues aside) that enrollees or employees comply with indications, avoid specific exposures, or take particular treatments to reduce the risk of illness, reduce health care costs, or improve quality outcomes. In the latter case, it is possible that state laws or an interpretation of the Americans with Disabilities Act could bar such action as illegal. The ethical question is whether such a requirement on the part of the state would constitute coercion rather than its right to protect its workers from self-imposed risk or harm (like the state's right to require everyone in a moving vehicle to wear a seat belt). Compounding this concern could be a situation in which the biomarker tested was later found to be associated with a far more deleterious health risk. In this case, the ethical ramifications could be great, particularly as any legislative efforts to maximize benefit while minimizing harm in these areas are intended to encompass the future. But if the risk/benefit ratio on which legislation was based changes dramatically for some, policy or law are challengeable on fairness grounds. Here, an individual's preference for genetic ignorance could later become a burden for the state, though the latter would not necessarily ethically justify a state's right to enforce mandatory testing.

Future Concerns

Well before the human genome project was completed, researchers and ethicists identified the need for privacy protections for genetic data banks. Protecting privacy is increasingly difficult as competition for research samples is keen and the ability to link disparate databases is ever growing. Protection is challenged by conflicts that can exist between common law, research ethics policy, and health information legislation.[22] As the need for new knowledge increases, so too does the importance of large population data banks. New information technologies are being designed to maintain data security, to improve data analysis, and to generate useful information for the end user, whether they be practitioners or consumers. The increasing global harmonization of ethical principles and standards raises important questions about the adequacy of such unification in both overcoming cultural and economic disparities and complying with original goals set forth by United Nations Educational, Scientific, and Cultural Organization and others international bodies in their issuance of concepts, principles, and terms to ensure that human genetic advances benefit all humankind (such as but not limited to the International Bar Association's Model Human Genome Treaty established in 1997) and subsequently established national policies.[23]

Conclusion

Fundamentally, the interests of investigators and participants are not in conflict. Ideally, the interests of companies offering genetic tests and solutions, and the end users who are practitioners and consumers are not in conflict. To maintain this commonality at all levels, explicit conditions of ethical acceptability must exist at all phases and levels of research as well as in commercial usage. Not-for-profit organizations have established themselves as credentialing agencies for different aspects of fair practice in the marketplace—for example, the Health on the Net code of conduct (HONcode) and TRUSTe licensure of privacy standards, and URAC,[24] whose mission is to promote health care quality through accreditation certification and other quality improvement activities. None of these organizations, however, are specifically focused on genetics and its integration into health care. Oversight by professional organizations that promulgate policies and practice standards, and the institutions that implement those guidelines, can help to ensure that the use of biomarker research and storage meets the highest ethical standards. Balancing protections for institutions seeking to further medical science and individuals and their families who want to advance genetics, but not incur adversity as a result of the participation, will be an ongoing effort. Ultimately, public confidence in the ethics and integrity of biomarker research, its translation into clinical practice, and the supporting commercial enterprises are vital to the success of the field.

Notes

1. Dr. Willard Centerwall discovered that applying a solution of ferric chloride to a wet diaper would produce a green color for babies with undetected PKU. The "diaper test" became the standard method of diagnosis in 1957.

2. Biomarkers are proteins, gene variants, or other biochemicals that indicate or can measure a biological process. Detecting biomarkers specific to disease can aid in the identification of risk, diagnosis, or treatment of affected individuals who do not exhibit symptoms.

3. There is no federal mandate for newborn screening. Each state determines its own testing policy and procedures.

4. Maryland is that one state.

5. Deidentified means that all personal identifying information linking a specimen to an individual had been removed, and the sample is referred to by coded identifiers only. For an overview of issues, see Skloot, Rebecca, "Body Stuff Politics: Taking the Least of You," *New York Times Magazine*, April 16, 2006, 38–81.

6. *Moore v. the Regents of the University of California*, 793 P.2d 479 (Cal. 1990).

7. Organs and tissues harvested at the time of death as a donation are subsequently processed (i.e., tested for disease and infection and sold as medical resources).

8. Religious beliefs are viewed by some as justification for exception.

9. For clinical research to meet conditions of ethical acceptability, it must not only meet conditions of informed consent but represent a valid scientific method as well.

10. See the Association of Accreditation of Human Research Protection Programs, http://www.aahrpp.org.

11. Mishra, R., "Harvard Is Asked to Tighten Rules in Clinical Trials," *Boston Globe*, May 13, 2004, A1.

12. Mertz, J. F., et al., "Protecting Subjects' Interests in Genetics Research," *American Journal of Human Genetics* 70, no. 4 (April 2002): 965–71.

13. Maienschein, J., "Editorial: Science Literacy," *Science* 281, no. 5379 (August 14, 1998): 917.

14. Mertz, J. F., et al., "Protecting Subjects' Interests in Genetics Research," *American Journal of Human Genetics* 70, no. 4 (April 2002): 965–71; Maloney, D. M., "Complaint Alleges Many Deficiencies in IRB-Approved Informed Consent Form," *Human Research Report* 20, no. 10 (October 2005): 8–18; Maloney, D. M., "Case Study: Agency Says Institutional Review Board (IRB) Failed to Warn Subjects of Significant Problems," *Human Research Report* 21, no. 3 (March 2006): 6–7; Nowak, K. S., Bankert, E. A., and Nelson, R. M., "Reforming the Oversight of Multi-Site Clinical Research: A Review of Two Possible Solutions," *Accountability in Research* 13, no. 1 (January–March 2006):11–24; Mertz, J. F., Magnus, D., Cho, M. K., and Caplan, A. L., "Protecting Subjects' Interests in Genetics Research," *American Journal of Human Genetics* 71, no. 1 (July 2002): 215; Santarlasci, B., et al., "Heterogeneity in the Evaluation of Observational Studies by Italian Ethics Committees," *Pharmacy World and Science* 27, no. 1 (February 2005): 2–3; Goldman, J., and Katz, M., "Inconsistency in Institutional Review Boards," *Journal of the American Medical Association* 248, no. 2 (July 9, 1982): 197–202; Prentice, E., and Antonson, D., "A Protocol Review Guide to Reduce IRB Inconsistency," *IRB* 9, no. 1 (January–February 1987): 9–11.

15. Skloot, Rebecca, "Body-Stuff Politics," *New York Times Magazine,* April 16, 2006, 38–81.

16. Washington University sued Catalona in August 2003.

17. www.house.gov/list/press/ny08_nadler/DOJAbor_021304.html; www.daily northwestern.com/news/2004/02/16/Campus/Feds-Subpeona.Nu.For.Abortion.Rec ords-1913789.shtml; query.nytimes.com/gst/fullpage.html?sec=health&res=9C0DE6 DF103AF931A25751C0A9629C8B63.

18. Rosenstock, L., and Landrigan, P., "Occupational Health: The Intersection Between Clinical Medicine and Public Health," *Annual Review of Public Health* 7 (May 1986): 337–56; see also http://www.orpha.net/data/patho/GB/uk-CBD.pdf.

19. *Toxic Law Reporter* 20, no. 13 (March 31, 2005); Beryllium Help Center http://bankrupt.com/CAR_Pub/0141.mbx; www.braytonlaw.com/practice areas/beryllium.htm; www.chronicberylliumdisease.com/news/nw_010204_environ. htm.

20. See www.atsdr.cdc.gov/tfacts4.html.

21. The Employment Retirement Income Security Act of 1974 sets minimum standards for pension plans in private industry. In general, ERISA does not cover government or self-insured group health plans.

22. http://www.ornl.gov/sci/techresources/Human_Genome/resource/privacy/privacy1.html; Caulfield, T., "Tissue Banking, Patient Rights, and Confidentiality: Tensions in Law and Policy," *Medicine and Law* 23, no. 1 (2004): 39–49; Mertz, J. F., et al., "Protecting Subjects' Interests in Genetics Research," *American Journal of Human Genetics* 70, no. 4 (April 2002): 965–71.

23. "Data Storage and DNA Banking for Biomedical Research: Technical, Social, and Ethical Issues," *European Journal of Human Genetics* 11, suppl. 2 (December 2003): S8–10.

24. See http://www.hon.ch; http://www.truste.org; www.urac.org.

CHAPTER 7

Genetically Modified Foods: Do We Become What We Eat?

If you are thinking a year ahead—plant seeds, If you are thinking 10 years ahead—plant a tree, If you are thinking 100 years ahead—educate the people.

—Kuan-Tzi, 4th–3rd century B.C., China

Scientists claim to have bioengineered a gene from a tiny worm that could one day lead to fatty foods that protect your arteries, not clog them.

—*Associated Press*, January 2004[1]

Foods are rich cultural mediums enabling connections to our ethnic heritages and reaffirming important aspects of who we are individually and communally. As such, food is packed with symbolic meaning. It can be associated with identity (bratwurst as "German" food), ritual (turkey on Thanksgiving, birthday cake), healing (chicken soup), and danger (poisonous mushrooms). Our relationship with food is greater than just the ingestion of nutrients and this is clearly evident in the diverse and vociferous opinions about genetically modified (GM) foods. The ethical controversies surrounding GM food involve many more ethical concerns than just the question of whether we ought to genetically modify food. GM food can be viewed as challenging culturally entrenched meanings. For some, the very existence of GM food represents a deliberate tainting of food as well as its cultural norms. In reality, a core point is that nature itself has genetically modified plant and animal life through survival. This means that, technically, we eat GM foods all the time. The relevant issue is whether the modification occurred naturally or resulted from deliberate human intervention.

Food production, regardless of whether it involves genetic modification, involves a wide range of biological, environmental, socioeconomic, and ethical issues. The need to feed the growing world population, especially the millions of starving individuals in pockets around the globe, is enormous. That demand is heightened by agricultural practices that damage the environment, such as soil erosion, chemical pollution, and reduced biodiversity. Particularly today, with increased demand for organic produce and chemically free meat, the ways in which food is processed and marketed arouses strong public sentiment. Add the use of cutting-edge technologies, such as genetic modification, to improve the quality and quantity of the food supply, and at least in Europe, citizens are quite concerned about not only the relative health risks and benefits but also broader normative issues, such as whether we have the right to genetically modify life, in this case, plant and animal. Animal welfare activists have for years objected to the exploitation of animals, such as choice of feed, chemical injections, and slaughter methods solely to please human palates or to increase profits for livestock produces. Opponents argue that these methods improve the safety of our food and efficiencies in its production, while reducing animal suffering. Moral debate about food production and distribution is neither new nor restricted to the adoption of genetic modification. One might think that GM food, with its possibilities of increasing the nutrient value of food, extending shelf live, and even improving efficiencies in production, and thereby offering an improved strategy for reducing famine and starvation, might create a moral imperative. In support of this view, consider that Norman Borlaugh (1914–), motivated by chronic food shortages in India and Pakistan in the 1960s, created a dwarf strain of wheat that increased yields by 70 percent and was able to use it to saving *a billion* lives in developing countries. For this work, he won a Nobel Peace Prize in 1970. Yet, GM food is politically contentious and ethically challenging both intrinsically as well as in application—namely, regarding human safety, access, and right to know.

The area of GM/GE (genetically engineered) food raises complicated ethical issues on global, societal, and individual levels. Among the goals of this chapter is to illustrate some of the many ethical issues and help the reader learn how to identify underlying ethical positions and distinguish them from strong emotion.

What Is Genetically Modified Food?

Biotechnology, a term that was coined in the early 1970s refers to techniques applied to living organisms, or their components, to make end products, such as wine, cheese, beer, or yogurt. Genetic modification uses the same principle as biotechnology, but it's more precise. Genetic engineering or modification refers to a set of specific techniques that alter the genetic makeup of a living organism, such as a plant, animal, or bacteria (see Table 7.1). Techniques are used to introduce, enhance, or delete a particular characteristic in an organism by altering the organism's DNA (genetic makeup), and as such, the application of the techniques is thought to take much of the guesswork out of the breeding process. That is one of several reasons why these techniques are supported. Older

Table 7.1
How Are GM Foods Created?

The process of genetically modifying an organism is complex. In general the process works this way.
1. Identify which gene(s) are involved in producing the desired trait.
2. Isolate the gene using DNA probes that find and splice off the fragment of DNA responsible for the desired trait.
   —The probe finds the piece of DNA that is desired, usually by using radioactive tags that bind to the gene and light up, permitting the gene to be identified and decoded.
   —The gene is cut at the appropriate place by restriction enzymes that recognize specific DNA sequences. Amplification of specific DNA sequence, and insertion of specific sequence into bacterial genome using DNA ligase enzymes.
3. The gene is then inserted into the organism in various ways, such as:
   —agro bacterium-mediated plant transformation
   —electroporation, which uses high-voltage current to make cell membranes permeable to allow the introduction of new DNA (commonly used in recombinant DNA technology)
   —microprojectile gun shooting of the DNA into the cells using a blast of high-pressure helium gas
4. Once the foreign DNA is successfully implanted, the plant cells are grown in the laboratory.
5. If the plants grow successfully, their seeds are harvested and then planted under field trial conditions.
6. If field trials are successful and approved, then the new seeds are commercially available for farmers and consumers to buy and plant.

techniques involve interbreeding of species (typically based on phenotypic properties) with *the hope* of obtaining the desired result.

Modifications at the molecular level typically involve changing one of the thousands of genes that exist within a particular organism. All living cells (plants and animals) contain molecules that encode the instructions for its growth and development. These molecules, of either DNA or RNA, are organized into sequences known are genes. Genes contain areas that are translated into RNA and others that are spliced, the function of which is still not completely understood. Cells have nuclei with their genetic information encoded in DNA, in which the DNA is transcribed to a messenger molecule (mRNA), which exits the nuclei and is translated by intracellular organelles to make proteins consisting of amino acids. These proteins, in turn, affect growth and development. Such a minute change can have profound effects. Scientists could insert into corn, for example, a gene from a common soil bacterium, *Bacillus thuringiensis,*[2] to confer resistance to damage from insect pests. Or, scientists might insert an animal gene into a plant to create new functions for the organism. For example, a gene that enables coldwater fish to survive, and even thrive, at low temperatures can be inserted into a plant to enable it to be cold-tolerant or frost-resistant. Such an organism is called a *transgenic organism.* Transgenic organisms are those produced by inserting genes from one species into another. In sum, molecular biological technologies can alter any food crop, animal product, or food ingredient during its production (see Appendix B).

The goal of GM/GE foods is to improve the nature of the food by increasing its resistance to disease, toxic herbicides, and insecticides; fortifying its nutritional value; and increasing its survival time (i.e., shelf life). Below are examples of existing GM foods; one plant based and one animal based.

- Plants are typically modified to resist frost, to yield higher protein and nutrient levels, to produce healthier oils, and to have a longer shelf life. For example, soybeans have been modified to resist herbicides that would normally kill them.

- Animals (including pigs, cows, and chickens) are typically modified for faster growth rates, leaner muscle mass-to-fat ratios, greater resistance to disease, and in the case of cows, to produce milk that contains a human breast milk component.

Like all food, GM foods are regulated by the Food and Drug Administration (FDA) for safety. All require premarket approval before being made available to the public.

A Brief History of GM Food

Breeding plants and animals for desirable characteristics is a practice that dates back centuries. Since ancient times, farmers stored seeds from the best-yielding and most disease-resistant crops, which were used for sowing the next spring. At its most basic, manipulating living organisms to yield specific traits or by-products, involves using a type of stimulus that "fiddles" with certain characteristics of a cell or cells. In other words, GM foods existed before the identification of DNA and the discovery of techniques to modify the genome of particular foods.

Fermentation, one type of manipulation, often referred to as a primitive form of biotechnology, was used as long ago as 6,000 B.C. by the Babylonians, Egyptians, and Sumerians to make beer and wines. The Egyptians are known to have made leavened bread by keeping a bit of sourdough (fermented bread dough) and adding it to new batches of dough. Molds were also used then to make cheese. Pre-Columbian farmers are noted for having been sophisticated cultivators with their techniques later adopted by the Incas to create the famous terraces of Machu Picchu in Peru, where more than a thousand different varieties of ancient crops have been found. Cross-breeding is another type of modification that has been used for the purpose of making a better crop. Cross-breeding, or cross-pollination, uses interspecies and intergenus crosses to produce food that can better adapt to climate, soil type, and other growing conditions. Hybridization, another technique, has been used for thousands of years to create crops more perfectly suited to human taste, if not also need. A London nurseryman, Thomas Fairchild (1667–1729) produced Europe's first interspecific (between species) flower around 1710, thus developing the first hybrid plant. The nectarines available today are a hybrid of oranges and peaches.

The process of cross-fertilizing was first discovered in 1724. Gregor Mendel (1822–1884) first described the inheritance patterns of specific traits of living organisms in his seminal experiments to create hybrids. Though used in some form thousands of years earlier, fermentation is considered to have originated on the basis of Louis Pasteur's (1822–1895) discovery that life can exist without oxygen (anaerobic life) and that under such conditions microorganisms—specifically what is now known as yeast—ferment. This discovery established the foundation for the field of microbiology. Pasteur found that microbes, such as fungus now known as yeast, can cause fermentation. In the late 1800s, cotton was crossbred to create hundreds of varieties with superior qualities. At the onset of the twentieth century, Luther Burbank (1849–1926) developed several new hybrid fruits as well as the Russell Burbank potato.

Heirloom varieties (including vegetables, grains, fruits, and flowers) permit us to get the tastes, feels, and appearances of plants of yesteryear. Some heirlooms are direct descendents of crops brought to the New World from Europe and beyond, while those native to the Americas have traveled around the world and back again. Heirlooms garnered great attention in the 1970s when small groups of growers became alarmed at the growing death of available heirloom variety seeds in favor of F1 hybrids from commercial growers. Their interest in preserving genetic variety spurred them on to collect traditional Indian seeds as well as those handed down from one generation of Amish and Mennonite gardens. Nonprofit organizations such as Seed Savers Exchange and Native Seeds/SEARCH now offer these seeds. An important benefit touted by heirloom varieties is the reinfusion of genes. Recurring hybridization shrinks the gene pool, because it not only narrows the species' pool, but also increases vulnerability of the breeding process to disease, climate consequences, and pest infestation.

In 1954, James Watson (1928–), Francis Crick (1916–2004), Maurice Wilkins (1916–2004), and Rosalind Franklin (1920–1958) identified the structure of DNA, the so-called blueprints for life. The discovery of double-helix and nucleotide strands, inherent in all plant and animal life, enabled scientists to understand how the individual nucleotides (A, T, C, and G) bind together and how specific traits are encoded for by the specific sequences of A, T, C, and G and then organized into specific genes containing exons (the coding regions) and introns (the noncoding regions). Building on this knowledge, scientists developed an understanding of how altering an organism's DNA could result in modifying that organism's traits. The identification of DNA and a greater understanding of its properties clearly paved the way for advances in recombinant DNA technology, which is the science of combining genes from different organisms. The organism resulting from the recombination is said to be *genetically modified, transgenic,* or *genetically engineered.*

The exploration of recombinant DNA techniques began in the early 1970s. Stanley Cohen and Herbert Boyer are thought to have launched modern biotechnology in 1973 with their success at inserting one gene of one organism into another. Boyer is credited with being the first to create a synthetic version of human insulin. Hence, there is a need to clarify the difference between genetic modification and genetic engineering, which refers to the process by which a genome is altered. The potential risk that genetic engineering represents is in the product(s) produced and *not* in the actual technique by which the modification has occurred.

Media reporting about these advances, and the potential utility-related applications, prompted public worry about "Franken foods" and represented the public's instant reaction to such reports about these techniques and their applications. In part because of this fear surrounding unknowns and uncertainties about biotechnology, the secretary of the U.S. Department of Health, Education and Welfare, in 1974, chartered the Recombinant DNA Advisory Committee (RAC), which was accountable to the director of the National Institutes of Health (NIH). The secretary charged it with the responsibility of developing regulations for recombinant DNA research, primarily as applied to human research, in part to reduce the potential for unwanted germ-line alterations. By 1976, the first federal guidelines for research were released, but they were revised in 1978 to relax many of the initial requirements.

Before 1980, it was impossible to patent any living organism, including microorganisms, because they were considered "products of nature" and thus not possibly patentable. Manipulating the DNA of organisms, however, changed that paradigm. In 1980, the U.S. Supreme ruled in the case of *Diamond v. Chakrabarty*[3] that microorganisms produced by manmade techniques were indeed patentable on the grounds that such products were created by human manipulation and, similar to any other invention, were made for a specific purpose. This decision was based on two legal precedents: (1) since 1930, certain asexually reproduced plants were afforded protection under patenting law; and (2) the 1970 Plant Variety Protection Act provided protection for some sexually reproduced plants. In fact, the Court ruled that genetically altered life forms *would require* patenting. The Court's decision permitted an oil company to patent an oil-eating microorganism. The ruling was highly significant, because it ushered in an acceptance of exploiting genetic engineering for commercial purposes and thereby opened the door for biotechnology to become big business.

In 1988, the U.S. Patent Office permitted the patenting of genetically manipulated animals. The first such animal, a mouse, was engineered to be highly susceptible to cancer and was created to test potential carcinogens. Since then, the technological capabilities have expanded to the point that the long predicted ability to change one species into another is now a reality, thanks to the newly emerging field of synthetic biology. New life forms, in particular microbes, that previously did not existed can be and are being created to perform specific tasks, such as producing biofuel or absorbing atmospheric carbon dioxide to eliminate global warming, generating chemicals needed for antimalarial drugs, and pumping out chemicals needed for a novel nylon-like fiber.[4] The issues and controversies surrounding the patenting of "manmade animal life" are highly complex and cannot be discussed at length here. Nonetheless, the potential dangers, extending well beyond the capacity for bioterrorism, have been acknowledged by at least 38 civil society and policy organizations that are pushing for greater government oversight, as the field seems to be outpacing the ethical, legal, and national security assessments. Given the confines of this chapter, only fundamental ethical questions will be taken up here.

The Availability of GM Foods

Today, GM crops are grown in field trials and commercially in more than 40 countries and on 6 continents. According to the U.S. Department of Energy's Human Genome Project Web site,[5] in 2000, a total of 109.2 million acres were planted with transgenic crops, principally herbicide- and insecticide-resistant soybeans, corn, cotton, and canola. In that same year, the countries producing 99 percent of the world's transgenic crops were the United States (68 percent), Argentina (23 percent), Canada (7 percent), and China (1 percent). Other crops include a sweet potato resistant to a virus lethal enough to decimate most of the African harvest, a type of rice with increased iron and vitamins ("Golden Rice") designed to help stave off chronic malnutrition in Asian countries, and a variety of plants able to survive at climatic extremes. As of 2002, 5.5 million farmers worldwide were said to be growing 50 million hectares of GM foods. Britain, however, does not permit commercial growing of GM foods.

Among the GM/GE foods now in field trials are bananas that produce human vaccines against infectious diseases, such as hepatitis B; fish that mature faster; fruit and nut trees that ripen earlier; and plants that produce plastics with unique properties. Some experts expect the production of GM foods to plateau in developed countries but increase rapidly in developing countries. Although identifying genes for important traits remains a significant limiting step, genome-sequencing programs for thousands of different organisms are yielding vast increases in knowledge that are likely to spur even more developments. Other GM foods available or currently in the pipeline include soybean, potato, pea, lettuce, maize, papaya, canola, sugarcane, pineapple, apple, grapevine, lentil, and wheat. Table 7.2 provides more examples and some of the improvements in these genetically modified foods and methodology.

Table 7.2
Intended Benefits in Designed Genetic Modification

Environmental Benefits	Examples
Less chemical pesticides and lower levels of carbon dioxide in the atmosphere	Herbicide-tolerant plants, Bt[a]-maize, rice with increased photosynthesis due to presence of maize genes
Increased precision and speed	Gene-transfer vectors and viruses, breeding methods, electropration, biolistics
Improved nutrition	Iron-enriched rice, Vitamin A-enriched rice ("Golden Rice")
Increased yields	Boosted rice output with maize genes

[a]Bt refers to proteins extracted from bacterium (*Bacillus thuringiensis*) present naturally in the soil worldwide. Biotechnology has enabled the more efficient use of Bt proteins to control insects that are harmful to crops. In other words, Bt technology reduces the need for recourse to conventional insecticides.

Source: Web-based curriculum for Balwyn High School in Victoria, Australia, http://www.balwynhs.vic.edu; http://www.bioportfolio.co.uk/news/agbio.shtml.

Separating Beliefs About GM Foods from the Ethical Issues They Raise

Screaming "GM foods are great" is not an argument for the ethical defensibility of GM food. The scream clearly represents an ardent, albeit vociferous, belief. But the ardor of that belief does not in itself morally sanction the existence or use of GM food. Even if the majority of people on the earth shared the same believe, that would not in and of itself demonstrate the morality of GM food. Whether something is ethical or not does not, as discussed in chapter 2, depend on belief. However passionately in favor or opposed to GM foods one might be, the strength of the emotions does not have a direct bearing on the question of whether it is ethical.

In addressing questions of whether genetic modification is ethical, we must understand the question posed and examine the reasons provided in support or refutation. In trying to analyze the question of whether GM food is a fundamental good or bad thing, we can begin by identifying criteria or standards of intrinsic "goodness" and "badness." To do this, we can apply various moral theories (as discussed in chapter 2) espousing rules for determining whether or not something is ethical. *Deontological theories* assert that there are many factors other than the mere goodness or badness of an act's consequences. For example, certain features of the act itself play a role in determining whether the act is ethical. Deontologists claim that an action can be morally right or even obligatory, even if it does not produce the greatest amount of good over evil, for self, society, or the universe. *Teleological theories* claim that the moral worth (whether something is ethical) is determined by whether the thing or act produces more good over evil (e.g., more pleasure over pain) than any alternative.

Throughout history, moral theories have concerned themselves far more with questions of duty than questions of altruism, although with more contemporary ethical theories, such as care-based ethics, we see a change in this direction. Ethical altruism, as discussed in chapter 2, is relevant here because one could argue that ethical altruism requires ensuring that all persons have enough to eat, even if it means denying ourselves some for the sake of others. A fundamental ethical issue surrounding GM food is whether we have an obligation, based either on duty or altruism, to feed the world. Former microbiologists and bioethics activist, Garrett Hardin (1915–2003) argued that we do not have such an obligation, writing that "it would be foolish for rich nations to share their surplus with poor nations, whether through a World Food Bank, the export of technology, or unrestricted immigration" because in light of the growing population of the developing world, such sharing would "do no good—it would only overload the environment and lead to demands for still greater assistance in the future."[10] In my opinion, his position is indefensible, but I invite the reader to formulate your own position taking into account the following factors.

In reality, all sorts of moral relationships exist between people. Many moral relationships are relatively straightforward, though the theoretical basis or justification for the relationship may be open to philosophical debate. Some relationships are less fundamentally straightforward. Consider that within the past few decades, many ethicists have said that we have a moral obligation to distant peoples and even future generations. Most of these ethicists, however, are not burdened with the many hypothetical considerations that such types of obligations raise. Police officers, for example, are obligated to help people in distress. Food manufacturers and processors are obligated to do no harm to those who consume their products. Though seemingly straightforward, such relationships get tricky in the context of ascertaining one's duty to others, particularly when the duty to the other creates a risk or liability to one's own self-interest. Consider issues that arise in the following hypothetical situation. Say 30 adults at an apartment complex's pool witnessed the drowning of a

7-year-old boy. None of the observers do anything to try to rescue the boy, and he drowns. Arguably, his life might have been saved if even one of those people acted. Morally abhorrent as it may seem to many, most laws do not prosecute failure to be a "good Samaritan." Separate and apart from whether neglecting starvation is an actionable offense, do we, individually or collectively, in light of our shared humanity have a moral obligation to act to prevent starvation? What about suffering? What if this obligation creates some level of personal or societal risk to our best interests? Applying the various moral theories to this and related questions will generate different answers based on different reasoned priorities. Regardless, the ethical defensibility of any position will rest on the soundness of argumentation and not on the ardor of one's passions.

Ethical concerns apply to the general question of whether we ought to genetically modify plant and animal life as well as to the myriad of questions raised by risk-/benefit predictions. Some claim that tampering with nature by mixing genes is absolutely wrong, no exceptions. The position further argues that mixing genes from different species is universally immoral, as doing so violates an organism's intrinsic worth and constitutes an affront to the shared dignity of life. Animal rights advocates, not surprisingly, object to inserting animal elements into plants, regardless of possible benefit. Other more general concerns focus on primary issues of rights and responsibilities to one another as human beings. Global benefit sharing is a vital ethical issue, for example, because poorer countries are unlikely to be able to afford the technologies from which they could benefit. Other ethical concerns are raised by more pragmatic considerations such as the public's right to know whether or not their food has been genetically modified, and hence the need for rigorous food labeling that indicates accurate details about any alterations that are contained in produce or ingredients.

Current Ethical Concerns as Evolving from Past Breeding Norms

As previously discussed, breeding to obtain desirable traits has been in practice for thousands of years. The crossbreeding of plants to produce hybrids of fruits and vegetables is valued because they have traits that we like, such as flavor, color, or texture. The breeding of animals to produce desired characteristics has been similarly considered highly successful and desirable, at least by some. Horses are bred for work, racing, or recreation. Dogs are bred for their desired characteristics such as work, show, and domestication. Much of our existing food supply is the result of market-driven forces that researchers and producers strive to meet. Most people, for example, prefer the white meat of the turkey. So, turkeys are now bred to have a shorter bone structure with heavier breast muscles, with the "coincidental" feature that they fit more easily into our ovens. Although breeding has become common practice, and even taken for granted, the fact that it may be customary and has been practiced for centuries does not in itself justify it as an ethically defensible practice.

Breeding, like GM food, implies that plant and animal resources are ours to do with as we wish. A fundamental ethical question here is whether we are entitled to break the fundamental genetic composition of species or even to break barriers between species to create transgenic life forms, particularly for our use or pleasure. The issue is whether, and if so, why, natural resources are ours to control. Some have begun to answer this question by theorizing the moral status of all living things, including animals and plants. It's not a stretch to maintain that all living things have equal moral status, but it could be a stretch to maintain that animals, plants, and other nonhuman living organisms can assert their moral agency or even interests. Some, on the other hand, object to the notion that all living nonhuman organisms have equal moral status on the grounds that animals and plants have inherent value, only less so than humans. Lacking reason, autonomy, and higher intellect, they (unlike humans) do not have moral agency. From this perspective, humans are considered to have the highest inherent worth, given their capacity for moral agency. Notably, this argument entails that the mentally retarded have less inherent value than "normal" people, because their mental "disability" compromises their moral agency, which in turn makes them less than fully able to exercise free will and accept responsibility for their actions. Such arguments create slippery slopes, for they suggest that a particular criterion, say intellect, is the defining criterion of inherent value. Following this line of argumentation, do geniuses have a greater inherent value than average folk like you and me? The logic here is less than fully compelling. Consider the question of whether humans have a right, or a duty, to use natural resources for their own purposes in a different context. Many people happily kill what they consider to be nasty mosquitoes, preventing at least the itchy mosquito bites or at best lethal disease, like malaria, while not believing that such actions have any moral status whatsoever. For some it's ethically acceptable to kill pests but not pets. For others, the willful killing of any living organism is immoral. The ethical justification of any position is only as good as the reasons argued. The following discussion raises and analyzes some of the many specific ethical questions surrounding GM food.

Do We Have an Obligation to Ensure That All Humans Have Access to Food?

Concern about the scarcity of food given the world's growing population dates back to the eighteenth century when Reverend T. Robert Malthus (1766–1834) cautioned us to control population growth lest this growth threaten our ample water and food supply, not just for human beings but all living things.[6] Mathus's concern for our planet's ability to sustain an increasing world population and all forms of life remains pressing in the twenty-first century. How does this apply to GM/GE food? One might argue that while everyone has a right to adequate food, we have an obligation to do our best to provide adequate food to those in need. The Food and Agriculture Organization of the United Nations (FAO), the World Health Organization (WHO), and other organizations report facts (from many sources) that would seem to compel such an obligation. In other words, what right do we have to let people starve to death, particularly if we're gorging ourselves on luxurious foods while they watch?

The current world population is estimated to be 6,315,872,951, and this number has grown at rate of 1.6 percent over the past two decades, averaging a yearly increase in the population of roughly 75,318,861.[7] It is generally agreed on that 2 billion people are currently undernourished. This means that roughly one-third of the world's population are malnourished; an estimated 1 billion people are severely malnourished, and another billion are said to suffer from microdeficiencies in their diet. A majority of these people live in the developing world. Millions of them are under the age of 5; 91 out of 1,000 children die before their fifth birthday in the developing countries compared with just 8 out of 1,000 children in the United States.[8] An FAO report indicated, however, that although the world's population is expected to reach 8 billion by 2030, the growth in agricultural production is expected to adequately meet the populations' food needs. Though the report projects a basis for optimism, it states that poverty and poor food distribution would continue to limit food access in some countries, thereby mitigating overall confidence in meeting the populations' food demand.

There are a number of other relevant statistics. It has been reported that 2 billion people worldwide live "below the bread line" (lacking the basic necessities of life and basic nutrients to sustain life) in developing countries. Some 500 million people, most of whom are under the age of 5, are predicted to not survive to childbearing age because they don't have enough food to eat. The U.S. Agency for International Development (USAID) predicted that if food production was organized as it is in Holland, there would be enough food to feed 67 billion people, which is 15 times the world's population. Food First, an international organization whose mission is to champion the right of every human being to food,[9] claims that there is enough food to feed everyone in this world: 4.5 pounds of food per person, per day. The problem is not one of production, but of access and distribution. Seventy-eight percent of countries that report child malnourishment export food.[10]

The British government pays out £26 million a year to farmers to stop them from growing food.[11] If 0.5 percent of the world's spending on weapons was diverted to agriculture in Africa, then three-quarters of that continent's poverty would be lifted. Americans spend 11 percent of their incomes on food while most of Africa spends 80 percent. The WHO claims the biggest killer in the world today is not cancer or any other disease but "deep poverty." In the United States, the richest society in the entirety of human history, 32 million people were living below the poverty line in 1988 (this was at the height of a 1980s economic boom) and nearly one in five children were born into poverty. Catholic Charities claimed that the number of meals they provided nationwide in 1996 had increased by 16 percent, and nights people spent in shelters has increased 35 percent. In Britain, one in three children grow up in poverty (defined as being half the average wage), and one in five households have no breadwinner. In 1992, the total economic output of the whole world was five times what it was in 1950, yet poverty is worse than what it was 45 years ago.

GM/GE foods offer the promise of meeting some of these food supply challenges. Only by breeding crops to resist potent crop killers (e.g., specific plant diseases, drought, pests, or other adverse factors) or to contain added nutrients, some argue, will we be able to efficiently and effectively feed the starving populace throughout the developing world.

The FAO declared that ethical aspects of GM/GE foods fall within the Universal Declaration of Human Rights. In 1996 and 2002, the World Food Summit and the Rome Declaration on World Food Security reaffirmed the right of every human being to adequate food.[12] What this right to food means, according to the United Nation's Committee on Economic, Social and Cultural Rights, is that—

The availability of food in a quantity and quality sufficient to satisfy the dietary needs of individuals, free from adverse substances, and acceptable within a given culture.... The accessibility of such food in ways that are sustainable and that do not interfere with the enjoyment of other human rights.[13]

The Special Rapporteur of the Sub-Commission on the Promotion and Protection of Human Rights of the United Nations Commission on Human Rights further declared that—

State obligations require active protection against other, more assertive or aggressive subjects—more powerful economic interests, such as protection against fraud, against unethical behavior in trade and contractual relations, against the marketing and dumping of hazardous or dangerous products. This protective function of the State is widely used and is the most important aspect of State obligations with regard to economic, social, and cultural rights, similar to the role of the State as protector of civil and political rights.[14]

Declaring access to adequate food a basic human entitlement is not the same as arguing that we have an obligation to provide everyone with adequate food, or that we have a duty to prevent famine, starvation, and malnourishment, that is, to provide adequate food to those in need.

Furthermore, justifying such an ethical claim is independent from enforcing it. Some argue that contextual and enforceability issues could reasonably mitigate a universal duty. For example, under ideal situations in which food could be both safely provided and optimally utilized by those who are in need, a universal obligation is compelling. Conversely, if ensuring that the quality and safety of the food provided was problematic (for all sorts of political, economic, and sociocultural reasons), one can counterargue that such a duty is mitigated by these circumstances because they limit, if not prevent, the ability to fulfill that duty. Consider these ethical questions: Am I less obligated if I know that my efforts will be foiled by any number of events? If the probability of succeeding is less, or even considerably less, than 100 percent, when does the duty hold and when does it not? At what point does the duty or "doing the right thing" reflect foolhardiness or unreasonable behavior?

Fairness: Are All People Entitled to Equal Access to the World's Scarce Resources? Is Equitable Access Possible in a Free-Market Economy?

The claim that we have an obligation to provide adequate food to the underfed and nonfed entails other claims that on face value require justification. In particular, the claim that we have such an obligation entails a premise that every human being has an equal right to an equal share of the world's resources. Concomitant to that claim is an argument that no single person or group of persons has a right to destroy, waste, or use more than a "fair share" of the resources. A notable characteristic of this line of argument is that it's headed for a debate on both ideological and pragmatic grounds. Ethicists on the opposite political spectrum argue that people in rich countries who allow or even promote the suffering and death of those in poor countries are committing "reckless homicide."[15] While it may be politically naïve to think that we could equally divide wealth and resources among all peoples, let alone ensure the maintenance of such equality, the self-interest justifying free-market economies are not without their liabilities and risks, if not morally indefensible processes and outcomes.

Fairness issues in setting policy priorities are always contentious, at least politically. More recently, the need for reasonableness and accountability has become a key requirement for establishing an ethically appropriate and justifiable framework for public policy debate and ultimately policy establishment. In the past, stakeholders with the greatest power and wealth set the agenda and policy priorities. While arguably that remains the case today, there are initiatives to provide greater "influential equity" among stakeholders. In addition to fairly balancing present needs by way of the introduction and commercialization of GM/GE crops that favor the environment, considerable attention must also be given to future matters, such as policies to ensure that the use of GM/GE crops is consistent with agronomic practices that nurture crop diversity, sound crop rotation, soil fertility, wildlife biodiversity, and the overall reduction of the adverse impact of agriculture on the environment.

Is Modifying the Genes in Food Ethical?

These questions can be approached in several ways. Among the simplest is to first consider what constitutes an ethical concern and then consider whether GM/GE food fits the criterion. The value of this approach is that it illustrates that an occurrence, event, or action is not ethical because it is either (1) popular, that is, it is the view held by those in power; or (2) the word of a divine entity understood by a chosen few. When we ask whether GM/GE foods are ethical, in one sense, we are asking whether GM/GE foods are intrinsically a good thing. Are they good in and of themselves, and are they good independent of the many uses of the food? We are also asking whether GM food is always intrinsically a good thing, never intrinsically a good thing, or sometimes intrinsically a good thing and sometimes not intrinsically a good thing, where the conditions supporting whether it's a good or bad thing require justification.

To GM or Not To GM

As stated earlier, the world population currently stands at roughly 6.2 billion, and given the present growth rate, estimates indicate that there will be another 2 billion people to feed in two decades. Feeding the world is an urgent problem, especially in light of expected population increases. Those are staggering numbers that create not only ethical debate but also moral urgency. A diet diverse enough to meet nutritional needs as well as pleasant to eat would seem highly desirable as well as necessary to feed the starving and malnourished.

Yet, eliminating malnutrition in a sustainable fashion with available natural resources is regarded by some to be next to impossible. Some argue that resources are already at their limit and may be now declining. Good growing land is under increasing pressures from spreading urbanization and industrial farming (e.g., soybean crops) and much of the good farmland has been depleted. The result is that new farmland is hard to establish without encroaching on environmentally sensitive or protected areas. Moreover, these areas tend to be low yielding anyway.

Most surface water is already being used for farming and any inadequacy, in both water and land, suggests that more efficient use is needed. Compounding the problem is changing weather patterns that affect growing conditions. These factors explain in part why conventional breeding methods have been unable to produce higher-yielding more robust plants. In short, existing methods of food production are insufficient. We need to produce more and better food with fewer resources. Genetic modification and genetic engineering offer possible solutions that promise better yields (due to more efficient photosynthesis) with greater nutritional values:

- Plants that are more resilient to climate extremes and problem soils
- Plants that are resistant to diseases and pests
- Production that conserves existing resources by—
- Improving water efficiency and reducing water loss
- Producing plants that need less agrochemicals and utilize nitrogen in the air

- Removing salt from the soil
- Shortening the growing season
- Enlarging plant leaves to deprive weeds of sunlight
- Reducing the need for plowing, which degrades the soil

In addition, GM foods promise to reduce spoilage and wastage through slower maturing varieties; improve the quality of food by removing toxins and allergens; enrich food with increased iron and other nutrients; produce foods that contain medical benefits, such as vaccines; and produce "better foods," like meats that contain less fat while being more tender, potatoes that soak up less oil in cooking, and oils that contain fewer trans-fatty acids.

Trade-offs

As with any new technology, the idea of GM "free rides" (i.e., those that provide a benefit all the time for everyone) is a myth. GM foods and food ingredients afford many benefits, but they also pose risks both known and, as yet, unknown. Controversy about GM/GE foods involves a wide range of concerns, including but not limited to the following: (1) safety for humans and other organisms; (2) short- and long-term environmental safety and the need for biodiversity; (3) adequate labeling; (4) informed consumer choice; and (5) intellectual property and patenting of life substances. Advocates and critics cite a wide range of benefits and detriments.

Touted Benefits

Beneficial aspects of GM foods include an increased food supply; the elimination of food-borne disease by engineering foods resistant to endemic viruses and bacteria; improved nutritional value of staple crops; and enhanced medical benefits in food (nutra-therapeutics) such as vaccines in food (see Table 7.3).

Risks Associated with GM Organisms in Human Food

The associated risks are not so much related to the specific process of creating GM/GE foods but from the traits introduced during the process. Genetic engineering typically targets genes with the goal of introducing one or two traits or modifications. The FDA's assessment is primarily focused on safety concerns. One concern related to the use of

Table 7.3
Some Benefits of Genetic Modification

Organism	Benefit
Crops	Increased yields, better stress tolerance, and faster maturation time
	Improved taste, more nutrients
	Improved growing techniques and new products
	Improved resistance to diseases, pests, and herbicides
Animals	Increased resistance, hardiness, and feed efficiency
	Better yields of products such as meat, milk, and eggs
	Improved animal health and diagnostic methods
Environment	Conservation of resources—land, water, and energy
	Improved natural waste management
	"Friendly" bioherbicides and bioinsecticides
	Bioprocessing for forestry products and more efficient processing
Society	Increased secure food supply for growing populations

GM organisms is the risks they could pose to human health. For example, some GM foods currently contain antibiotic-resistant genes that may contribute to the spread of pathogenic bacterial strains that are resistant to currently available antimicrobial agents. Similar to this is the concern of using antibiotics to improve the yield from cattle, which could alter the natural gastrointestinal flora of humans and cause adverse effects. Another much discussed concern is the safety risk associated with the transfer of plant-derived transgenes to intestinal microflora, which is a highly complex ecosystem. The issue here is that genetically altered bacteria contained in food, when humanly ingested, may adhere to gastrointestinal epithelia and invade intestinal enterocytes, which could result in the transfer of genes from microorganisms to mammalian cells. According to a 2002 editorial in the *Lancet*,[16] many of the perceived risks have scientific merit. Health concerns include new allergies and other adverse human health effects resulting from gene transfers, particularly of antibiotic-resistant genes, from GM foods to cells in the gastrointestinal tract. There is concern for "autocrossing," which is the movement of genes from GM plants to conventional crops, that indirectly threatens food safety and security. Other touted adverse effects are GM crop threats to biodiversity; decreases in the richness of food, including but not limited to cross-pollination of GM organisms with "natural foods"; and farmer dependence on chemical and biotech

Table 7.4
Feared Risks of GM Foods

Safety, for humans as well as for the environment
Gene flow and the introduction of adverse consequences: introduction of new
  strains like "super weeds" or antibiotic resistant bacteria
Introduction of new allergens
Reduced biodiversity, and genetic uniformity
Pesticide-resistant insects and other pests that threaten the food supply
Threats to organic farming technology
Safety in labeling (presently not required in the United States)
Mix of GM with non-GM products or ingredients on labels confuses
  consumers
Societal concerns: new techniques and products will advantage the rich
Access and intellectual property
Biopiracy: exploitation of world's resources by a few
Domination of the world's food by only a few companies
Dependence of the developing world on the developed countries

companies, with attendant economic and political effects felt globally. Some of these threats are noted in Table 7.4.

The Public, Farmers, and Governments Weigh In

The views of different stakeholders—farmers, consumers, and producers—do not change the ethical issues, but they do provide valuable insights and perspectives that are bound to grease or "grittify" acceptance of policy and regulatory decisions. For example, the term "genetic modification" can suggest something that is highly technical, not easily comprehensible, and therefore to be feared. Similarly, the term "Franken foods" is used by activists opposed to GM foods to convey how frightening and potentially unsafe the creation of new transgenic life forms may be. So-called Franken foods have been noted as causing fear in not only the public but farmers as well. A 2003 survey of Canadian farmers and the general public revealed that farmers fear "Franken foods" and most favor the labeling of GM/GE food. Many farmers throughout the developing world report concerns about the GM/GE controversy.[17] Understandably, their main concern is the impact on their income.[18] Notably, though, others oppose mandatory labeling based on the fear that sales will plummet, especially if labeling is required on all modifications, including cross-breeding and selection processes.[19]

The public has its own views. U.S. studies have reported that a large majority favored the labeling of GM food and indicated that they were willing to pay more to ensure that accurate labeling. Most people said they wanted access to detailed information about GM products. Despite a high level of trust in the FDA and its oversight of the safety of the food supply, a majority of those polled believed that regulations governing GM products were inadequate. Table 7.5 illustrates some trends in U.S. public opinion. Interestingly, most of the questions revolve around the issue of labeling despite the fact that studies show that people choose products on the basis of price and not labels. Of note is the fact that, despite apprehension and public ignorance about the presence of GM food in their diets, GM crops have been adopted at a high rate over the past decade.[20]

These surveys are interesting windows into underlying normative beliefs about GM food. While results show a consistent desire for the labeling of GM food, research indicates that current labeling doesn't really affect consumer behavior. The opinion polls in Table 7.5 are interesting also because of the questions *not* asked. A sampling of recent polls, for example, does not include questions about consumer concerns about biodiversity, ecological safety, or the potential for an adverse impact on developing countries, nor do they solicit opinions about the numerous and varied ethical issues facing humans, other species, and the environment that arise in the use of GM food.

Can GM Food Be Deployed in Ethically Appropriate Ways?

If we were to agree, at least for the sake of argument, that (1) we have an obligation to do our best to reduce famine, starvation, and malnourishment; and (2) we ought to further develop and use GM foods as an efficient and effective strategy for achieving that goal, there nonetheless remain innumerable ethical issues associated with the use of GM foods. The FAO recently stated that "major changes in the fields of food and agriculture in recent years, including accelerating technological development, changes in the resource base, and economic and market developments, have brought to the fore a variety of ethical questions of relevance to food security and sustainable rural development." For this reason, "Ethics in Food and Agriculture" was designated as a priority area for interdisciplinary action and, under Article VI.4 of the Constitution, a panel of experts in this area was established along with an internal FAO committee to guide the organization's actions.[21]

Table 7.5
A Decade of American Public Opinion Polls about Food

Year	Source	Opinion
2004	Pew Initiative on Food and Biotechnology	—54% of Americans know little or nothing about genetically modified foods —89% say that no such food should be allowed on the market until the FDA determines that it is safe
2004	Rutgers University Food Policy Institute Study	—43% of Americans think (incorrectly) that ordinary tomatoes do not contain genes, while genetically modified tomatoes do —30% think (incorrectly) that eating genetically modified fruit would change their own genes
2003	ABC News Poll	—55% of Americans would avoid genetically modified foods if labeled —one-third of the population already tries to avoid food that is genetically engineered or treated with hormones or antibiotics —62% of women, who do the lion's share of grocery shopping, would avoid genetically modified foods —an overwhelming majority of Americans want genetically modified foods labeled —46% report that they believe that genetically modified foods are unsafe
2001	ABC News Poll	—52% of Americans report that they believe that genetically modified foods are unsafe —93% say that the federal government should require labels, whether genetically modified or bioengineered
2001	Farm Foundation Survey	—90% of farmers say that biotech products should be labeled if scientifically different from conventional foods
2000	MSNBC Live Vote	—86% of Americans want labels on genetically engineered food —89% think the government should require pre-market safety tests of genetically engineered food

(Continued)

Table 7.5 (*Continued*)

2001	Pew Initiative on Food and Biotechnology	—75% of Americans say that it is important to them to know whether a food product contains genetically modified ingredients
2001	Rutgers University Food Policy Institute Study	—90% of Americans say that genetically engineered foods should be labeled
2000	*USA Today*	—79% of Americans say that it should be illegal to sell genetically modified fruits and vegetables
1999	Edelman Public Relations Worldwide	—70% of American consumers want more extensive labeling of genetically engineered food ingredients
1998	*Time*	—93% of women want all genetically engineered food labeled
1995	*Food R&D*	—94% of consumers believe that milk should be labeled to distinguish milk from rbGH-treated cows —36,000 people polled say that they want genetically engineered food labeling: 94% of women are pro-labeling, 84% of men are pro-labeling
1994	Center for Food Safety	—88% favor mandatory labeling of rbGH-treated cows and products derived from such cows, 9% oppose mandatory labeling, and 3% are unsure
1989	Center for Food Safety	—77% of North Carolinians feel producing more nutritious food is the best use of genetic engineering of food —80% of those polled say that genetically modified foods pose serious health risks to humans —67% of those polled believe that genetically modified foods will give large farmers an unfair advantage over small-scale farmers

Source: Center for Food Safety.

Determining whether the genetic modification of food—if ethically managed—can be a genuinely positive step toward providing adequate nutrition to those desperately in need involves assessing a manifold of operational issues, including but not limited to the following: seed cost;

the cost of other inputs that the farmers will have to use; the cost of cultivation; and the potential for market price increases, which arguably subject those who are already in need to more hunger and starvation. The ethics of operationalizing GM food in a global free-market economy run deep, indeed. Cost-benefit ratios may well be the most crucial factor, as the ability and the moral commitment of the well-fed to provide to the needy is pivotal.

Consumer Right to Know versus Industry Rights to Avoid Undue Burdens

A consumer's right to know is one focal point of concern about GM/GE foods. The right to informed choice, which derives from the ethical concept of individual autonomy, is key to ensuring that the use of GM foods is ethically sanctioned. Just as we want to know which medications have adverse side effects and which cars have the highest crash safety rating, we want to know how our food is produced, as well as the associated risks and benefits of that production process. The "right to know" about your food involves numerous issues and considerations, not the least of which are the complexity of information to be disseminated, the likelihood of comprehensibility by the mass population, and the financial cost of converting the information into an easily understood and regulated format.

The right to know how food is produced, particularly whether it is organic, "natural," or genetically modified, as well as the full content of particular ingredients, has become increasingly important to consumers who seek to comply with increasingly more specific nutritional advice. Consumers must be able to know what they are consuming to make informed choices, so full disclosure in labeling is crucial to upholding this "right to know." But what constitutes full or adequate disclosure?

Food labeling, already laden with numerous ethical issues (not the least of which is a producer's legal, regulatory, and ethical obligation to inform consumers so that they can make informed choices), has become even more contentious as consumers demand to know potential risks and benefits. Included in labeling obligations is the importance of informing the consumers of potentially significant harms. These concerns are relevant on a local and an international level.

Ensuring that consumers are fully informed is a burden shared by consumers (who must educate themselves enough to understand what GM food is and how to interpret food labels) and the industry (who must provide disclosures required by law or regulation). Industry compliance with law and regulation can be costly, and given primary obligations to their shareholders, industry typically seeks to avoid adverse impacts on profitability, thus pitting consumers' right to know against industry's right to profitability. But even if the industry provides full disclosure as required by regulation, how can consumers be sure that what's disclosed constitutes the whole picture? In other words, do non-disclosures of information significantly change the meaning and impact of what is disclosed?

Moreover, consider the responsibility of the consumer to educate her/himself about the health and safety issues involved in labeling. More important, informed choice and resulting actions require access to information and resources. Consumers do not all have the same access to information and resources to make informed decisions about genetically modified organisms (GMOs). Particularly in developing countries, the very poor (both women and men) may lack the most basic information to make decisions that may affect their health and their capacity to sustain themselves or—worse—they may lack the literacy required to understand that complete and accurate information is not present. Appropriate methods to reach the least educated, the poorest, and the most disadvantaged groups should form part of any strategy to inform the public so that individuals are able to choose according to their needs. This challenge is formidable. Consumer access to accurate information can be crucial, particularly if you have a metabolic disorder, food allergy, or food sensitivity. Individuals with hereditary hemochromatosis, for example, metabolically store excess iron, and these individuals should stay away from the many iron-enriched foods. In fact, individuals diagnosed with various metabolic conditions, or allergies, have lobbied for increased and more thorough labeling to ensure that they're not exposed to harmful ingredients. For individuals with acute food responses, knowing the full contents can literally be the difference between life and death. For example, the incidence and severity of peanut allergies has made the thoroughness of labeling essential.

Dietary and nutritional advice is abundant, but in the United States it rarely includes advice about eating GM food, particularly because U.S. labeling standards of GM foods are nearly nonexistent. Consumers, particularly in Europe, are demanding to know whether foods are the result of genetic modification. These same people aren't clamoring to know the full extent of the food's pedigree or breeding, nor even the specifics about the genetic technologies employed in creating modified foods. Rather, they simply want to know whether the product is or isn't determined to be a GM food.

This raises the question of whether GM labeling is more important than other types of food labeling. Would you, for example, want to know whether a particular food was genetically modified rather than the nutritional value of that food, or would you prefer to know whether it was vitamin enriched (e.g., labels stating iron fortification in cereals or calcium fortified juices)? Would a seal of endorsement by the American Heart Association make you more likely to buy one brand of orange juice over another? Or, what if you have an allergy to something in a GM food, but wouldn't know it until after you digested it and had an allergic reaction? Would that experience differ clinically from having an allergy to a hybrid fruit or vegetable, and if so, would that difference be important for treatment and management of your reaction? Why or why not?

The right to participate in policy setting and all aspects of decision making about GM/GE foods is important to ensure that the use of GMO is ethical for all affected. This right to democratic participation addresses the need for justice and equity, which is a major concern in the context of GMO-related decisions. Principles of justice may include gender equality, need, accountability, liability, and fair and democratic procedures. Many young people, particularly the poor and powerless, have little education and no social entry point to influence decisions about GMOs. They ought to be given every opportunity to participate in the debate concerning the impact GMOs may have on their lives and livelihoods, and the potential benefits that may arise from the development and use of such products. They should also have the right to choose the product that best suits their needs.

Another major concern is the fact that future generations have no voice or vote in current decisions about GMOs, which implies a current duty to establish mechanisms to ensure that future interests (potential interests) are adequately taken into account *today*. Strategies to meet anticipated and unanticipated future needs should be kept alive to allow for the handling of unanticipated future needs, such as those deriving from unpredictable environmental changes.

Conclusion

While not every new technology successfully entrenches itself in daily life, GM foods are not likely to go away. Their promised benefits are great, particularly in light of the urgent need to feed large sectors of the world population who are underfed and malnourished. The existence and use of GM food, while providing substantial needed benefits, may nonetheless result in unanticipated adverse effects despite our best efforts. Guarding against such effects may well be our greatest ethical challenge, particularly as we currently produce enough food to feed the world. Starvation and malnourishment result from failures of distributive justice that bar access to food. Access and just distribution of food to all humans is an urgent problem not limited to GM foods, but rather to all food (genetically modified or not). As such, it is particularly important from an ethical standpoint to ensure that the benefits of GM foods and it risks are shared equitably among all, and not disproportionately allocated in ways that perpetuate the vulnerable bearing great burdens, resulting in malnourishment and even death due to starvation.

Advances in genetics and genomics are revolutionizing food production, as well as nutritional science, with the new fields nutrigenomics and metabolomics providing insight into how genetic makeup affects nutrient response. Nutrigenomics is showing us that not only do our nutritional needs vary according to our genetic makeup but that nutrients can affect gene expression, and thus reduce disease risks. Feeding ourselves what our genes require for optimal functioning, in other words, can mitigate against predispositions to health risks. One strategy to achieve both the reduction (if not the elimination) of starvation and malnourishment and optimal health through customized diets based on individual nutritional needs is to create optimal foods through the marriage of genetic modification of food and nutrigenomics. In other words, knowing an individual's optimal nutritional requirements could spur development of new foods and beverages as treatments or preventive agents for individuals, families, or subgroups predisposed to a particular disease.

These technological capabilities raise ethical concerns not only about whether and to what extent we should adopt them but, if so, how to ensure the appropriate application (including global benefit sharing), given that the same genotype may not confer the same risk in all populations. Among serious ethical concerns is the fact that some of the gene variants associated with health risks that *can* be mitigated through improved diet may also be associated with other disease risks for which no treatment or prevention is available, thus creating conflicts about the value of knowing and not knowing one's genotype. Conversely, uncertainties about the long-term safety and health concerns about GM foods and how they may affect gene expression remain.

Apart from these concerns are other ethical questions, such as whether we ought to produce food that meets the needs of various groups based on their biological requirements, even if such production is barely profitable. If, or really when, the field of nutrigenomics advances to the point that we can test everyone to determine their unique metabolic rates to stratify nutritional requirements based on metabolic profiles, should population screening and tailored foods be supplied, analogous to personalized medicine, which is designed to ensure that the right drug is given to the right patient in the right dosage? Clearly, once we start considering application issues, we're immediately confronted with a cascade of ethical issues that are not restricted to disparities in access to food.

Notes

1. Kang, J. X., et al., "Transgenic Mice: Fat-1 Mice Convert n-6 to n-3 Fatty Acids," *Nature* 427, no 6974 (February 5, 2004): 504.

2. Bt refers to proteins extracted from bacterium (bacillus thuringiensis), which is present naturally in the soil worldwide. Biotechnology has enabled the more efficient use of Bt proteins to control insects that are harmful to crops. In other words, Bt technology reduces the need for recourse to conventional insecticides.

3. laws.findlaw.com/us/447/303.html; web.mit.edu/invent/iow/chakrabarty.html.

4. Lartigue, Carole, "Genome Transplantation in Bacteria: Changing One Species to Another," *Science Online*, DOI 10.1126/science.1144622, June 28, 2007.

5. Gautam, Naik, "J. Craig Venter's Next Big Goal: Creating New Life," *Wall Street Journal*, June 29, 2007, A1–2.

6. http://www.ornl.gov/sci/techresources/Human_Genome/elsi/gmfood.shtml.

7. Malthus, T. R., *An Essay on the Principle of Population as It Affects the Future Improvement of Society; with Remarks on the Speculations of Mr. Godwin, M. Condorcet, and Other Writers* (London: St. Paul's Church-Yard, 1798); Malthus, T. R., *An Essay on the Principle of Population: A View of Its Past and Present Effects on Human Happiness; with an Inquiry into Our Prospects Respecting the Future Removal or Mitigation of the Evils Which It Occasions*, 6th ed. (London: John Murray, 1826).

8. U.S. Census Bureau, "Total Midyear Population for the World: 1950–2050," http://www.census.gov/ipc/www/worldpop.html.

9. World Hunger Year, "Domestic Hunger and Policy Facts," http://www.worldhungeryear.org/info_center/just_facts.asp.

10. http://www.foodfirst.org; *Washington Post*, September 15, 2000, A26.

11. India and poverty, http://www.foodfirst.org.

12. www.eufactsfigures.com, October 2002.

13. FAO Corporate Document Repository, "GMOs and Human Rights: The Right to Adequate Food," http://www.fao.org/DOCREP/003/X9602E/x9602e03.htm#P0_0.

14. General Comment 12, paragraph 8 (E/C.12/1999/5).

15. www.un.org/rights/ (E/CN.4/Sub.2/1999/12).

16. Singer, P., "The Famine Relief Argument," in *Morality in Practice*, ed. J. Sterba (Belmont, CA: Wadsworth Press, 1988).

17. Rosenstein, D. L., "How Safe Is GM Food?" *Lancet* (2002): 360.

18. http://sunflowerstrewn.wordpress.com.

19. "World: Developing Countries Threatened by US Farm Bill," June 19, 2002; See also "USA/Zimbabwe: Food Aid Rejected Because it includes GM Corn-Members," August 2, 2002, www.Just-Food.com; Iowa State University, Biosafety Institute for Genetically Modified Agricultural Products, May 2002, "GAO: Testing of GM Foods Is Adequate, Monitoring of Health Risks Not Needed," http://www.bigmap.iastate.edu/www/285.htm.

20. Iowa State University, Biosafety Institute for Genetically Modified Agricultural Products, May 2002, "GAO: Testing of GM Foods Is Adequate, Monitoring of Health Risks Not Needed," http://www.bigmap.iastate.edu/www/285.htm.

21. Genetically Engineered Organisms Public Issues Education (GEO-PIE) Project, "GE Foods in the Market," Cornell Cooperative Extension, 2003, http://www.geo-pie.cornell.edu/ (accessed October 2003); Hallman, W. K., Adelaja, A. O., Schilling, B. J., and Lang, J. T., "Public Perceptions of Genetically Modified Foods: Americans Know Not What They Eat" (New Brunswick, NJ: Food Policy Institute, Cook College, Rutgers–The State University of New Jersey, 2002); Hallman, W. K., Hebden, W. C., Aquino, H. L., Cuite, C. L., and Lang, J. T., "Public Perceptions of Genetically Modified Foods: A National Study of American Knowledge and Opinion," Publication number RR 1003–004 (New Brunswick, NJ: Food Policy Institute, Cook College, Rutgers–The State University of New Jersey, 2003); Hallman, W. K., Hebden, W. C., Cuite, C. L., Aquino, H. L., and Lang, J. T., "Americans and GM Food: Knowledge, Opinion, and Interest in 2004," Publication number RR-1104-007 (New Brunswick, NJ: Food Policy Institute, Cook College, Rutgers–The State University of New Jersey, 2004).

CHAPTER 8

Cloning and Stem Cells: Fact, Fiction, and Ethics

Matteo Alacran asks El Patrón's bodyguard, "How old am I? ... I know I don't have a birthday like humans, but I was born." "You were harvested," Tam Lin reminds him.... To most people, Matt is not a boy, but a beast.... But for El Patrón, lord of a country called Opium—Matt is a guarantee of eternal life.

—Nancy Farmer, *The House of the Scorpion*[1]

Cloning: In Fact and Fiction

The quotation above is an excerpt from a popular contemporary novel, one of the many science-fiction explorations of cloning over the past 50 years. Matt is cloned from a powerful yet aging lord, El Patrón, to be a genetically identical source of replacement organs. As a clone (cloning, in this chapter, refers to the duplication of DNA fragments [small pieces of the DNA sequence], individual cells, and entire organisms),[2] young Matt becomes increasing self-aware of being an entity that is both human and distinctly not, and what it means to live in a society in which clones are despised, treated as livestock, and routinely slaughtered for their body parts. The tale challenges us to debate whether cloning threatens to transform an essence of humanity into material comodification.

February 2007 marked the 10-year anniversary of the first mammalian clone created by using genetic material from an adult cell (i.e., lamb

#6LL3, named "Dolly") that was generated by somatic cell nuclear transfer, using a differentiated cell from a mammary gland of an adult ewe. The headline that ran on February 27, 1997, "Scientists clone adult sheep-triumph of UK raises alarm over human use" referred to an announcement by Ian Wilmut (at the Roslin Institute near Edinburgh) and his colleagues (at PPL Therapeutics in East Lothian) that they had produced Dolly who had been born the previous July.[3] Although Dolly arguably shocked the world, the reality is that scientists had been developing techniques to reproduce plant and animal cells for decades. They knew for more than a quarter of century, for example, that adult skin cells from frogs could support the development of tadpoles when transplanted into oocytes (immature ovum),[4] but whether the same could be true for a mammalian cell wasn't known until the birth of Dolly. That press release in February 1997 announcing that a baby lamb had been born that was successfully cloned from an adult ewe[5] represented an answer to a basic question of developmental biology, namely "Does a differentiated adult cell retain the capability to direct the development of an entire adult individual?" The answer was now known to be "yes," under certain conditions.

The newfound ability to successfully clone mammalian cells heralded in new ethical concerns about the moral acceptability, or defensibility, of applying cloning techniques, which have led to laws banning human cloning.[6] In the past 10 years, what cloning is has been more precisely defined on the basis of methods, conditions, and anticipated result, or type of cloned entity. Ethical, if not also legal debate, has focused on the general moral acceptability of any cloning (for some any distinctions are pseudo distinctions). For example, some argue that cloning for therapeutic purposes (meaning cloning for the purpose of developing new treatments, if not cures, for diseases, particularly insidious ones such as multiple sclerosis or Parkinson's disease) is morally justified, if not imperative. Conversely, others vehemently oppose any form of cloning as a moral affront to human dignity and therefore feel that it is absolutely unjustifiable. Notably, however, ethical debate, launched by Dolly and encouraged by science-fiction stories, has changed over the past decade. What didn't happen was the birth of a cloned child or widespread public demand for the use of cloning for human reproduction. Instead the debate is far more complex, rooted in the reality of scientific research, including a merging of the debate into the sphere of embryonic stem cell research.

Nonetheless, and as is typical with demonstrated technologic advances, many of the questions about whether we should clone surfaced only *after* the appearance of Dolly, not before. Knowing that we can use cloning clearly raises deep questions about whether we ought to. And ought we at all? Or for some purposes but not others? And soon we are immersed in a quagmire of ethical concerns. Science, however, is way in front of moral debate, which raises its own ethical concern. Should science continue unchecked, because it can demonstrate what is and isn't feasible and thereby clearly frame our ethical concerns? Or should we to decide as a society, or even as a global community, whether or not science should proceed? Around the globe, different countries have taken different types of action,[7] with the United States most notably having banned federal funding for stem cell research, at least until June 2007.

This chapter reviews the historical background leading up to Dolly's birth, with a concomitant focus on cloning that is not restricted to human uses of molecular biological techniques, because examples exist in nature. Demarcating ethical acceptability from unacceptability, in this arena, entails sophisticated argumentation along a slippery slope of what is and what could be.

Embryonic Stem Cells Can Be a Source of Clones

The news media, including scientific journals, are replete with controversy about stem cells and their use. Stem cells are simply precursor cells, where the daughter cells may differentiate into other cell types. In pre-embryonic development, the *zygote* (fertilized egg) undergoes a series of cell divisions that apportions the large zygote into many smaller cells called *blastomeres*. These blastomere cells form a small ball named the *morula*. The morula further develops an outer and inner layer of cells with a fluid-filled center; the inner cell layer is not evenly distributed but lies against one side of the outer ring. This collection of cells is known as the *inner cell mass of the blastocyst,* and it is the origin of the embryo proper. Embryonic stem cells are derived from the inner cell mass of the blastocyst. They have the ability to develop into any of the more than 200 cell types that are found in human beings. As such, these cells are termed **totipotent**. Isolation of these cells destroys the embryo proper.

Adult stem cells are derived from a small population of stem cells in differentiated tissues. These cells still maintain a limited potential to develop into different cells types, but, in contrast to embryonic stem cells, adult stem cells can't develop into the full panoply of cells and tissues that comprise the human body. Adult stem cells are **multipotent**; they are restricted to developing into specific subpopulations of cell types. For research to develop these capabilities, stem cell lines need to be differentiated from the more generic cell lines, which refer to a population of cells propagated in culture that are cloned from and, therefore, genetically identical to a single common ancestor cell.

The primary ethical issue concerns the fact that isolation of these potentially valuable stem cells involves the destruction of the embryo. A further ethical issue concerns the fact that precursor cells can be manipulated to differentiate into nearly every cell and tissue type in mammals, which inspires visionaries to dream of their potential curative properties for diseases that have plagued humankind (such as diabetes, birth defects, and cancers) and their healing potential for disabling physical injuries, such as spinal cord injuries. Despite their promise, some fear that such manipulations amounts to unjustifiable tampering with nature. Some hold deep religious objections to all such research, including the degradation of what they define as human life. Some argue that if the embryos have been "discarded" (i.e., miscarriages as opposed to abortions), then research is more ethically acceptable, as abortions represent the willful destruction of human life. Others, who do not view embryos as already human, based on the nonexistence of sentient life or other characteristics of at least newborn status, do not view research on embryos as human degradation. Furthermore, some argue that even if one were to accept embryonic research as in some way debasing human life, the potential for such research to develop treatments, even cures, for human beings is a greater good that far outweighs any objections at the embryonic level.

In August 2001, President George W. Bush announced a ban on the use of federal funds for the development of further cell lines beyond the group of 60 cell lines available as of that day: "No federal funds will be used for: (1) the derivation or use of stem cell lines derived from newly destroyed embryos; (2) the creation of any human embryos for research purposes; or (3) the cloning of human embryos for any purpose." All federally funded stem cell research is now under the auspices of the National Institutes of Health (NIH) and is regulated by this ban. Private research, however, continues without this limitation. Moreover, as of this writing (June 2007), the president arguably bowing to political pressure issued a new executive order increasing federal funds to expand research on existing stem cell lines, where "the life and death decision has already been made. On June 20, 2007, President Bush announced:

> Earlier today, I issued an executive order to strengthen our nation's commitment to research on pluripotent stem cells. This order takes a number of important steps. The order directs the Department of Health and Human Services and the NIH to ensure that any human pluripotent stem

cell lines produced in ways that do not create, destroy, or harm human embryos will be eligible for federal funding. The order expands the NIH Embryonic Stem Cell registry to include all types of ethically produced human pluripotent stem cells. The order renames the registry—calls it this, the Pluripotent Stem Cell Registry—so it reflects what stem cells can do, instead of where they come from. The order invites scientists to work with the NIH, so we can add new ethically derived stem cell lines to the list of those eligible for federal funding. I direct Secretary Leavitt to conduct an assessment of what resources will be necessary to support this important new research.[8]

Background

Cloning in Molecular Biology

The cloning of DNA fragments, that is, specific sequences of nucleotides, is the backbone of recombinant DNA technology. By using restriction enzymes (which are derived from bacteria and used to cut DNA) that recognize a specific nucleotide base sequence and cleave that sequence from the larger DNA sequence, fragments of DNA can be isolated. Then, using DNA ligases, which are enzymes that catalyze the joining of two molecules to effectively link sequences of DNA together, the fragments can be inserted into a bacterial genome for subsequent amplification for in vivo cloning. Further growth of the "engineered" bacteria produces an abundance of the "engineered" DNA.

In vitro DNA fragments or genes can be amplified by use of the polymerase chain reaction (PCR). PCR allows amplification of a specific fragment of DNA by using repeated cycles of heating and cooling that separates helical chains of DNA, allowing primers for the sequences of interest to bind to the separated DNA chains. These chains form a soup containing single nucleotides followed by the addition of ligases that build complementary strands. The process is repeated over and over to produce a large quantity of specific fragments of DNA in specially built reactors.

As the molecular age arose so, too, did concomitant ethical concerns about this form of cloning, which was overseen by the NIH Recombinant DNA Advisory Committee.

Somatic Cell Nuclear Transfer Technique

Dolly's birth affirmatively demonstrated that a committed cell, that is, a cell that has a limited expression related to its differentiated state, still contains the instructions for the development of an entire individual. Scientists wanted to know whether it was possible to rewind the nucleus to a more primitive, *totipotent* stage. If so, one could reprogram to a more primitive, or totipotent, stage that could dictate the developmental programming for an entirely new organism. To succeed would be an important first step in realizing the curative potential in these cells. Some ethical concerns had to await the actual success of Dolly to be raised. Others gained greater urgency once Dolly's success was demonstrated. Still others await further scientific clarification, such as whether the clone will have an increased rate of senescence related to the origin of the nucleus, a differentiated cell. If so, then the success is mitigated and benefits of the techniques are considerably more limited.

As stated, the technique used to clone Dolly is called *somatic cell nuclear transfer*. By using an intact nucleus from a differentiated cell—in Dolly's case, a cell from an adult female sheep breast—disintegration of naked DNA was avoided. The intact nucleus was transplanted into an unucleated oocyte, providing the oocyte with its entire genetic composition other than mitochondrial genome that was present in the oocyte's cytoplasm. The success was limited. Many embryos didn't develop, and the success of clones coming to term was less than 2 percent. The reasons for the poor efficiency of the process are numerous, and many are related to the state of mammalian DNA in the zygote and the state of the cytoplasm of the enucleated oocyte.

Out of 227 eggs fertilized with this technique, only 30 began the first stages of embryonic division, and out of these, only 9 fertilized eggs caused pregnancy when placed in a primed surrogate ewe. Factors contributing to the low level of efficiency of this process need to be understood before cloning becomes a routine method of producing identical individual animals. And while many specific ethical concerns hinge on scientific success or failure, other more basic questions, such as whether we have the right to manipulate nature (animals) remain vital. Subquestions arise as well, such as whether we have the right to manipulate animals, particularly knowing that there is a high likelihood of creating malformed animals, who will suffer from their defects. Some of these questions are akin to the long-standing debate about animal rights and the ethical appropriateness of conducting experimentation on animals for human gain. Animal rights advocates, for example, have argued for decades that drug and cosmetic development testing on animals is simply unethical and that to preserve animal rights and dignity such testing should involve synthetic substances capable of yielding equally meaningful results. Whether animals should be bred merely for such testing is but one of several debates framing the starkness of the issues.

Can the Techniques Work for Human Cloning?

These unresolved questions hold the application of such techniques to human cloning in check. If applying such techniques to animals is ethically dubious, for example, how could it be possible to ethically justify researching these techniques in humans? While other scientific and medical questions raise reservations about the viability of cloning in humans, the overall ethical justifiability is far from proven or accepted. And notably, large concerns remain despite the relatively recent presidential ban seeking to quell debate, if not also research. Moreover, specific concerns will emerge on the basis of what science is able to demonstrate.

Will human clones who have received their genetic constitution from only one parent be more likely to accumulate deleterious mutations? We know that uniparental **disomy** is recognized as a source of human disease. In this situation, the loss of one parental gene occurs with duplication of the other parental gene to produce the normal 2N state. This phenomenon can explain the occurrence of recessive disorders, such as cystic fibrosis in which only one parent is a carrier.

Will clones have adequate DNA repair mechanisms that can correct somatic mutation errors of transcription? Normally, genetic mistakes are handled by one of three pathways: (1) they are recognized and repaired by DNA repair enzymes, (2) they are sloughed off by the process of cell death (apoptosis), or (3) the errors can reproduce indefinitely to the detriment and possible demise of the organism (e.g., cancer). Many questions surface about the long-term health of individuals with uniparental genetic inheritance. Would we be creating people who essentially have a time bomb ticking away inside of them? Such questions are clearly subsumed underneath overriding concerns about whether we ought to pursue these scientific questions. And if so, whether some types of scientific research are ethically permissible while others are not. And if so, why?

Why Clone Animals?

Cloning offers the possibility of replacing long-established methods of selective mating of adult animals for desired traits in their young as a means of animal breeding. Selective mating has been used for centuries to produce healthy offspring of prize-winning bulls, stallions, and other livestock and that over generations has attempted to improve the particular breed. Cloning offers a more efficient and exacting way to develop a herd of animals or a flock of fowl expressing a particular desirable trait. In the future, this could include animals bearing a human transgene. Ethical questions about the defensibility and appropriateness of these techniques include whether we have the right to meddle with the lives of animals in a way that only benefits our species. Consider some of the touted benefits and whether the advances carry greater moral weight than an outright ethical objection based on religious views such as those discussed earlier. Consider furthermore whether these advances are different in kind or degree from other types of advances, such as personalized medicine or behavioral genetics, which have their own ethical implications for human responsibility, obligations, and questions of justice. In light of all the many ways in which technologies and medical advances have improved human life, might these advances carry a moral imperative?

Being able to produce identical animals might prove extremely valuable in animal research as a way to eliminate diversity of biologic background, confounding the results of specific experiments. A variety of experiments on a single set of cloned animals might help researchers refine their search for mechanisms of disease and potentially discover different healing techniques, develop specific tissue for **xenotransplantation** (discussed below), or find more efficient methods to develop a set of desired traits in cloned offspring. Another real benefit of using cloning could be the development of a transgenic animal that expresses a particular desirable human gene that could then be cloned to develop new and efficient sources of useful drugs. For example, transgenic lambs have been developed to express human coagulation factor IX in their milk.[9] What is the impact of having a human protein circulating in the blood or milk of animals? How would it affect the animals supplying it? Is the question of whether we have the right to clone animals for our benefit any different than whether we have the right to eat animals?

Additionally, cloning offers the possibility of salvaging endangered species not only by increasing the number of animals available for breeding, but also by maintaining biodiversity that is optimal to support life for humans and for other plants and animals, arguably an urgent consideration in light of the ecological destruction caused by global warming. By making it possible to use genetic material from animals incapable of breeding, including very young animals and even dead animals, cloning can increase the pool of animals capable of reproduction. This, in turn, would decrease the chance of inbreeding with its associated risk of recessive disorders, given the higher carriers rate of deleterious gene alleles. In light of global warming and other environmental factors that threaten the viability of all life on the planet (e.g., the overuse of combustible gasses, deforestation and the resulting increased scarcity of animal feed, and other human behaviors that deplete environmental resources and ecological balance), might cloning be morally imperative as a means of ensuring life?

Cloning presents the possibility of enhancing the potential of xeno-transplantation to solve the tremendous backlog of patients awaiting a solid organ transplant. In xeno-transplantation, tissue from one species is transplanted into another. Pigs, for example, are being bred for the specific purpose of supplying heart valves that can be transplanted into human beings. Here, a pig's heart valves were genetically engineered to express compatible human antigens, which means that the proteins are recognized by the immune system as human rather than foreign, making the valves more readily accepted in transplantation. This lessens the transplant patients' requirement for immunosuppressive drugs, which, in turn, lessens the risk of the side effects the drugs cause. It is even conceivable that the patient's own embryonic stem cells preserved at birth from the umbilical chord can be the source of cells that could be directed down a developmental pathway of tissues needed for transplantation.

Obviously, demonstrating the feasibility and safety of these technique as well as resolving ethical issues is requisite to widespread adoption. However, even though cloning is currently fraught with risks and potentially negative results (as reported in the experience with Dolly, multiple congenital anomalies were seen in the unsuccessful products of pregnancy), the ethical question of whether we are justified in proceeding unless or until we have better scientific answers remains. For example, given the low rate of efficiency seen in reported studies thus far, ought research continue, or rather should it try harder? Is it justifiable to use a technique that has such high risks of failure? Presently, U.S. society has said "no" for proceeding with cloning in humans, and roughly 100 member countries of the United Nations voted similarly.[10] Might cloning of the so-called lower animal species continue, and if so why? Finally, how do we rectify the obvious ethical conflict in condoning cloning of some life forms but not others?

Possible Use of Reproductive Cloning in Humans

Reiterating the major caveats stated above, there are a number of situations in which cloning for human reproduction might be considered ethically defensible. For human diseases in which the maternal mitochondrial genome contains a faulty gene, the ability to use a surrogate or substitute oocyte (egg) with its healthy mitochondria and the couple's fertilized zygotic nuclei in vitro, using nuclear transplantation, might afford the development of a child without the risk of mitochondrial disease. In this situation, the nucleus transplanted would carry genetic material from both parents and hence would not be strictly an individual parental clone. The ability to use this technique raises the question of whether, from an ethical standpoint, it is morally different from pre-implantation diagnosis, in which case viable embryos are removed from the uterus and tested for genetic disease, and only those testing negative for genetic disease are reimplanted and grown to full term. Pre-implantation diagnosis is currently an acceptable practice to avoid primarily lethal genetic diseases and typically is used after a family has birthed a child with one such disease. Though comprehensive utilization statistics are not available, the technology is thought to be used primarily by the affluent, raising its own ethical concerns.

The combination of cloning with *gene therapy*, which is the introduction of a specific sequence of a gene to avoid disease associated with specific gene mutations, is another area in which cloning could be the therapy of choice in reproductive medicine. In the case in which one parent is a carrier of an autosomal dominant disorder, for example, Familial Adenomatous Polyposis (FAP), a condition of multiple colonic polyps that have a high likelihood of developing into a malignant tumor at a relatively early age. The parent who is free of disease could have cells cloned, so that the resulting offspring presumably will be free of the disease. It is unlikely that this strategy would be pursued, however, because most autosomal disorders have variable expression and the risk is not likely to justify the approach. An approach not involving cloning might be a scenario akin to the situation in mitochondrial disease, where an affected zygote, as detected by premimplantation diagnosis, could undergo gene therapy with **antisense nucleotides** to knock out the effect of the dominant mutated gene.

As a mean of assisted reproduction, cloning currently offers no advantages over in vitro fertilization (IVF) techniques. It is easy to imagine that cloning might be helpful to couples who are infertile and want a child whose is related genetically at least to one parent. Critics of human cloning bring up the issue of the singular identity of the clone, and the responsibility of the cloned individual to his or her clone. One response to that criticism is that the clone would not be totally identical. The zygotic cytoplasmic environment would not be identical to the parent, the uterine environment would be different, and the health of the original mother would be different than the cloned individual's mother. Once born, the clone would exist in its own time and space (not that of its origin) and, as such, its life experiences from embryonic stages through adulthood would be unique, much like it is for a natural child. Nonetheless, even this personal identity of a clone to itself is insufficient to prevent the commodification of life if the clone was used as a mere means to an end. Such use is morally abhorrent.

Cloning as the answer to the extreme shortage of solid organs for transplantation and thus a conduit to the preservation of life is a real consideration. Cloning has the potential to address both the shortage of donors and the issue of tissue incompatibility and thus could permit us to heal the ill and injured and maintain the lives of our loved ones. A newly created cloned embryo's inner cell mass could be harvested and directed to differentiate into the tissue of need for subsequent transplantation without growing into a full human being. Such a procedure, however, leads directly back to the controversial idea of developing embryos destined for destruction solely for the benefit of another person. Here the ethical issues overlap with the divisive question of precisely when an embryo attains personhood.

Do we have the right to create and then destroy an embryo solely for the purpose of saving another's life? Are we obligated to do everything we can to save human life? Whose life? Whose life is more important, the embryo's or the person's? In the early days of renal hemodialysis, sufficient numbers of machines were not available and hospital committees decided who was more worthy of dialysis. Those who did make the cut either received a less effective peritoneal dialysis or progressed along the lines of the natural history of their renal failure. It came down to whose life was worthier, and in those cases, perhaps the younger patient would be viewed as more deserving. Today, we still see the conflict between the pregnant woman and her fetus. From relatively benign procedures such as magnetic resonant imaging to invasive procedures, the fetus's right of protection (in the United States) frequently trumps the mother's rights. With the advent of assisted reproductive technologies, many fertilized embryos are in frozen storage awaiting a decision on the part of the biologic parents to their fate: continued storage at a cost or destruction. What is the current opinion as to the personhood of these

4, 8, or 16 cell embryos? When considering human cloning, some of the same ethical dilemmas that met IVF technologies reappear.

Clones versus Twins

Though exceedingly speculative, and with many countries having outlawed human cloning, it is nonetheless important to understand the difference, biologically and ethically, between clones and identical twins (nature's version of clones). Much has been written about the concerns that a clone's personal identity, and even autonomy, will be severely impaired by the fact of being a clone of someone else. In response, useful comparisons can be made to the natural development of identical twins. It is the division at the very early stage (4, 8, or 16 cell) that gives rise to identical twins. Would issues of autonomy be any different for clones than identical twins who share identical genetic constitutions? Such issues don't seem to plague nature's identical twins that share identical constitutions.

As time goes on, identical twins acquire somatic mutations that promote differences between them. Some physical features begin to differ and internal shifts occur that affect their physiology differently. Identical twins are never totally identical even though they share most of the same genetic material. Each has different exposures to DNA damage from sunlight, toxic substances, and mutations in DNA that occur with the duplication of cells as they age. Hence, twins lose some their genetic identity over the years of their lifetimes as they grow from neonate to elder.

Identical twins are also individuals who have different personalities, experiences, and responses to events; their personalities and mental characteristics may vary greatly. That variance is believed to be the product of gene and environmental interactions. That variance would be even greater in a clone who had not shared the same **oocyte cytoplasm**, the same womb, or the same nurturing environment. There may be special concerns about an asymmetric parent-child relationship because of the cloned child being so similar to his or her parent; however, this may be a mute point. The cloned child may simply be considered the same as his or her siblings who aren't clones and who share a genetic constitution with both parents.

Concerns About Reproductive Cloning

A primary concern is the safety of the technology. Is it possible that cloning may create a higher risk of birth defects than occur in natural reproduction? We should ask ourselves, however, what we mean by natural reproduction. The success rate for a successful implantation of a fertilized egg is not 100 percent. A number of pregnancies are lost in the first trimester, perhaps in the first week. Studies of these products of conception have shown a high rate of aneuploidy, an abnormal number of chromosomes. In any pregnancy, there is no 100 percent guarantee of a perfectly normal child. Couples are allowed to continue their attempts to become pregnant and increasingly use available technologies to optimize their desired outcome, whether it be preference for one sex over the other, identifying risk of genetic disease in utero, or even aborting fetuses diagnosed with Down syndrome or spina bifida. Furthermore, the rates of success with the current techniques of assisted reproduction are not 100 percent either. Though the actual rates vary with the individual practitioner, the general rates quoted for successful pregnancy are around 15 to 20 percent. Then there are the risks associated with prenatal procedures, such as amniocentesis and chorionic villus sampling. Current risk of loss of pregnancy with amniocentesis is 1 to 2 percent and 1 percent for chorionic villus sampling (CVS). Limb anomalies have been associated with CVS performed late in the pregnancy.

Before cloning would be permitted, an exhaustive study of the potential deleterious effects on the embryos would be required. However, such a large-scale clinical trial, similar to those that take place before a drug is permitted to be marketed, is not ethical and will never take place. A civilized society does not permit such a trial with such risks. This question arose during and immediately after World War II, and such a clinical trial would be in violation of the Nuremberg Code.[11]

More extensive study of IVF embryos for gene expression that correlates with successful implantation, health, and normal pregnancy, and that result in a viable outcome, could possibly take place using animal models. These studies would have to be reported in peer review journals and be replicated by multiple investigators. Some studies have already been completed on IVF embryos that have found that the expression of the insulin-growth factor family genes correlates with the growth potential of the embryo.

In the United States, it appears that the current ban on reproductive cloning, that is, cloning for the purposes of uterine implantation, is appropriate. According to the World Health Organization, cloning for human reproduction is ethically unacceptable and contrary to human dignity and integrity.[12] To date, while there has not been a U.S. federal statute prohibiting reproductive cloning, there have been multiple attempts in Congress since 1997, and numerous organizations, including but not limited to the Institute of Medicine have issued recommendations for a ban.[13] One of the reasons for Congressional inaction on this issue is that various bills offered have linked reproductive cloning and therapeutic cloning.[14] However, were it to be possible to learn a great deal about normal embryo development by using cloned embryos, the ethical question about whether bans should be lifted might well arise. Related ethical concerns involve questions about whether it is *ever* appropriate to develop embryos for research purposes using animal models. Would knowledge gained necessarily be transferable to humans? Nevertheless, the reality is that medical research proceeds from animal models to clinical trials.

Personal Choice versus Society's Choice

Other than safety concerns, ethical debate about reproductive cloning tends to center on questions of genetic determinism. We are more than a collection of our genes. As stated, if reproductive cloning is available, then the clone's oocyte cytoplasm environment would be different from the parent, interaction between the oocyte's mitochondrial genome and the clone's genome would be different, the intrauterine environment would be different, the health of the surrogate mother would be different, and perhaps the circumstances of the delivery would be different. All of these differences would occur before a clone's entrance into a postnatal world. The time difference, the circumstances of early developmental environmental differences, and a host of other conceivable environmental conditions would render the clone a unique individual despite the similarity of genetic material. Much like a natural twin, so many opportunities to differentiate will be confronted from the time of the genetic manipulation of stem cells that it would be impossible to be identical to the parent.

Exceptional circumstances have been offered as arguments against a complete ban against reproductive cloning, such as the death of a beloved child and for parents who are both carriers of a lethal gene. Even if the ban took into account such exceptions, another question looms. How would this practice be regulated, and how and who would determines the guidelines for exceptional use of cloning? It is unlikely that, even if reproductive cloning was permitted, many people would take advantage of it. One important factor is the cost. It would be expensive, and insurance probably would not cover the costs. It is also not clear how may people are really interested in genetic copies of themselves.

If or when cloning techniques have been perfected, ethical questions about appropriate applications, if any, will come to the fore. As discussed, the potential advancements delivered by types of genetic engineering are likely to once again raise the specter of eugenics. Hopefully, the sad legacy, at least in the United States, will serve to prevent such deterministic thinking and the abuses that would likely result. Cloning, if available, would open the door to a genetic underclass. Those for whom cloning would be unaffordable would have to resort to "natural" means of reproduction. Such offspring would miss out on genetic enhancements. A popular movie, *Gattaca* (1997) explores aspects of this genetic underclass. The hero was born without access to prenatal genetic screening and has to wear contact lens, which would exclude him from becoming an astronaut, but he manages to become an elite astronaut through sheer will.

The discussion to this point has focused mainly on cloning for reproductive uses and has given less emphasis on the uses of therapeutic cloning. There are arguments for and against therapeutic cloning. One argument for therapeutic cloning is its potential to answer the desperate need for solid organs. Statistics from the United Network for Organ Sharing (UNOS) show that more than 92,000 persons are waiting for an organ. An argument against cloning for organs is the fact that to extract blastomeres from a cloned embryos to develop immuno-compatabile organs, assuming we succeed in organotypic cloning, requires the destruction of an embryo. To create a cellular transplant, for instance, of **hepatocytes** or cardiac **myocytes**, one would isolate the blastomeres and direct them along the developmental pathway of either the liver or heart. Whether a cellular transplant would be effective to reverse the diseased organ is not known. Whether a true three-dimensional recreation of the organ would be required is the subject of additional research. What is known, however, is that even this mode of transplantation would require the destruction of an embryo. It is exactly this issue that opponents to cloning raise.

Does the created embryo have rights of its own? Where do federal and state rights to protect citizens come into play? From a social utilitarian perspective, it can be argued that, in the greater scheme of things, federal funding to pursue this line of research could be better spent on prevention of disease, so that the need for organ transplantations would decrease with control of such health risks as hypertension. Yet, consider how other ethical theories presented in chapter 2 would address this question. Again greater emphasis on preventive care might also lower the incidence of hypertension-associated cerebro-vascular disease, a lower incidence of renal failure related to the epidemic of obesity and obesity-related diabetes,[15] and better programs to treat drug addiction with a decrease in incidence of liver and kidney failure related to effects of intravenous drug addiction, and lower rates of AIDS, for example. From the deontological argument, would it be the right thing to do? What about from a care-based ethics principle? Who determines whether it is the right or wrong thing to do? On a global scale, should our focus be on this technology when there are still many people without clean drinking water, adequate food, and other necessities of life? Clearly, a fundamental ethical question is whether the development of cloning technology is a defensible use of available resources, and if so, who ought to have access and why?

Cloning raises several issues, like the other issues discussed in this book, that demonstrate the need for an informed global public. We must educate ourselves and others about genetic technologies, rationally define and discuss the ethical-moral issues that are entailed, and develop the analytical skills to assess for ourselves, for our communities, and our societies the best way to proceed.

Notes

1. Farmer, N., *The House of the Scorpion* (New York: Simon & Schuster Children's Publishing Division, 2002).

2. http://www.ornl.gov/sci/techresources/Human_Genome/elsi/cloning.shtml.

3. http://www.nature.com/news/2007/070219/full/445800a.html; Smith, L. C., and Wilmut, I., "Influence of Nuclear and Cytoplasmic Activity on the Development *in vivo* of Sheep Embryos after Nuclear Transplantation," *Biology of Reproduction* 40 (1989): 1027–35; Roslin Institute, lead scientist, Professor Ian Wilmut and graduate students from Kings College, http://www.roslin.ac.uk/public/cloning (April 2005); Wilmut, I., "Dolly—Her Life and Legacy," *Cloning Stem Cells* 5, no. 2 (2003): 99–100; Wilmut, I., "The Search for Cells That Heal: The Creator of Dolly the Cloned Sheep Asks That Society Look Past the Controversies to the Ultimate Payoff," *Scientific American* 293, no. 1 (July 2005): A35.

4. Gurdon, J. B., Laskey, R. A., and Reeves, O. R., "The Developmental Capacity of Nuclei Transplanted from Keratinized Skin Cells of Adult Frogs," *Journal of Embryology and Experimental Morphology* 34 (1975): 93–112.

5. Smith, L. C., and Wilmut, I., "Influence of Nuclear and Cytoplasmic Activity on the Development *in vivo* of Sheep Embryos after Nuclear Transplantation," *Biology of Reproduction* 40 (1989): 1027–35.

6. National Conference of State Legislatures, http://www.ncsl.org/pro grams/health/genetics/clone.htm.

7. www.advisorybodies.doh.gov.uk/uksci/global/japan.htm; mbbnet.umn.edu/scmap.html.

8. http://www.whitehouse.gov/news/releases/2007/06/20070620-8.html.

9. Schnieke, A. E., Kind, A. J., Richie, W. A., Mycock, K., Scott, A. R., Ritchie, M., Wilmut, I., Colman, A., and Campbell, K. H., "Human Factor XI Transgenic Sheep Produced by Transfer of Nuclei From Transfected Fetal Fibroblasts," *Science* 278 (1997): 2130–33.

10. www.geocities.com/giantfideli/art/CellNEWS_UN_cloning.html.

11. The Nuremberg Code was updated by the Declaration of Helsinki in 1964. See Appendix C for a list of the basic principles of ethical human experimentation.

12. World Health Organization Assembly, Resolution WHA51.

13. National Conference of State Legislatures, www.ncsl.org/programs/health/genetics/humancloning.htm.

14. Center for Genetics and Society, "Federal Policies on Cloning," http://www.genetics-and-society.org/policies/us/cloning.html#3a.

15. Bray, G. A., "Epidemiology, Trends, and Morbidities of Obesity and the Metabolic Syndrome," *Endocrine* 29, no. 1 (February 2006): 109–18.

CHAPTER 9

Fast Forward to the Future: Will Genetics Technologies Bite Back?

As discussed throughout the previous chapters, predicted impacts of genetic advances forecast substantial health benefits for individuals, their families, and societies. The chronological integration of new genetics and genomics knowledge and technologies was traced against the backdrop of regrettable misapplications. This perspective illustrates that "progress" is characterized by positive and negative, foreseen and unforeseen effects. Metaphors, like the double-edged sword, have been invoked to illustrate this duality of promise and peril. Implicit in prognostications about the impact of genomics, though often unacknowledged, is the reality that trade-offs are an integral component of every decision-making process. In actuality, progress is rarely, if ever, risk free, and actualizing the potential benefits of genetic advances, like other technological advances, is unlikely to be achieved without cost. Moreover, despite the allure of innovation, adoption (or integration into daily life) is never certain, because factors other than technological merit are involved.

As a society, we often turn to experts, and government, to guide us in determining which innovations are best to adopt, when they are good enough to be available for use, and how they can be appropriately integrated into daily to maximize benefit while reducing harm. Experience shows, however, that not only is the road to predicted adoption of innovation rocky (i.e., not linear but rather jagged) but that experts, whose advice we attend to in virtue of their superior knowledge and skill, are not necessarily good forecasters. If we look to the information technology (IT) industry, the point is quite aptly made by the following quotes:

> I think there is a world market for about 5 computers.
> —Thomas Watson, chairman, IBM (1943)

> There is no reason for any individual to have a computer in their home.
> —Ken Olson, president, Digital Equipment Co. (1977)

> 640K [of memory] ought to be enough for anybody.
> —Bill Gates, CEO, Microsoft (1981)

They missed the mark, but the creators of a certain mischievous cartoon character, Felix the Cat, got it dead right in 1962: "Home is where the computer is."

The point is that, while the future is likely to involve genomic technologies, the nature of that integration of genomics is not a given. Our future, like our genes, remains within our control to shape and realize. Appreciating that the integration of innovation invariably involves trade-offs and unforeseen adversities, so, too, does the integration of genomic technologies. Acknowledging this fact can prepare us to achieve more ethically defensible policy decisions that attempt to minimize unintended harms. In his book, *Why Things Bite Back: Technology and the Revenge of the Unintended Consequences*, Edward Tenner makes, in my view, a compelling case for the paradox of progress. Tenner offers the concept of "revenge effect." This is a distinctly different notion from unintended adverse consequences. Revenge effects, unlike, unintended adverse effects, are intrinsic reactions to the imposed solution, which serve to mitigate the overall improvement gained. Put differently, the impact of such revenge effects is to cancel out the benefit conferred by the technological solution. In so doing, a new and more insidious problem than the original one is generated. According to Tenner's theory, soon after the adoption of an innovation, the allure of limitless utility gives rise to an instability that inevitably converges to create some type of limit or vacuum elsewhere that nearly cancels out the "improvement." In other words, the revenge effect is nature's uncanny, and even ironic, way of "getting us back." Tenner writes,

> Whenever we try to take advantage of some new technology, we may discover that it induces behavior which appears to cancel out the very reason for it. The electronic gear that lets people work at home doesn't necessarily free them from the office; urgent network messages and faxes may arrive at all hours, tying them more closely to business than before.[1]

Analyses about the potential for adverse consequences resulting from implementing genomic solution strategies have as yet to consider the possible relevance of Tenner's theory. The purpose of this chapter is to do just that to shed light on possible normative implications for our "genetic" future. Its goal is to enhance sensitivity to the possibility of genetic revenge effects that, if we accept Tenner's notion, are inevitable inherent negative by-products of well-intended genomic solutions. The actual and possible (intended and unintended) ill effects, or, more properly, ethical injustices, resulting from misapplication and ignorance of genetic advances fit Tenner's definition of adverse consequences, not revenge effects. This chapter explores the possibility of as-yet-unknown revenge effects resulting from genomic advances. While Tenner would not necessarily support the claim that revenge effects can be identified in advance of their occurrence, the following discussion attempts to hypothetically explore this possibility. Inasmuch as we stand to benefit from the new capabilities spawned from genetic advances, one could argue that in virtue of our shared humanity we are obligated to strive to ensure that integrated genetic innovations are applied in ways that optimize benefits for all while reducing harm.

Applying Tenner's theory to genomic technologies suggests the possibility that nature will unleash revenge effects that will inevitably undermine the promised benefits of the genomic revolution. Determining what exactly Tenner might forecast is a bit tricky because of the need to distinguish adverse consequences, such as a genetic underclass, which have been analyzed and the basis for proactive prohibitive legislation and policy, from what would be a true revenge effect. The distinction, arguably, lies in the fact that the revenge is nature's bite-back effect, and one that is wholly independent of human doing.

What Drives the Adoption of Innovation?

Though it's a common belief that technology succeeds on its own merit, 75 percent of all new products fail in the marketplace. Why? Because usability, benefits, and end-user satisfaction are at odds with underlying values, beliefs, and problem-solving approaches (which can be drivers or barriers to adoption) such that the innovation poses a "cognitive misalignment" with the user. Technological capability and superior engineering, in other words, do not guarantee acceptance. Failures are attributable to many factors, including solutions for nonexistent problems. But underlying a product's failure, more often than not, is the designer's failure to understand the end user, namely, those particular values, beliefs, and attitudes that motivate behavior, that is, use. Building functionality to cohere with underlying cultural values is often the key variable in whether a technological innovation will succeed or fail. Factors integral to users' experiences, such as intuitive approaches to problem solving, environmental factors, or underlying beliefs, values, or attitudes play an *integral* role in adopting innovations.

The *truly* revolutionary technologies that succeed transform not just how we live but our culture as well. Take, for example, the phone answering machine, a technological advance that replaced human answering services. In the early adoption phase, when most people did not have services like voice mail, callers were startled when the phone was answered by a machine and not a real person. Some refused to leave messages, feeling indignation at having to talk to a machine. Gradually, as more and more people used answering machines, the culture began to change. Instead of indignation at having to talk to a machine, people began to find value in being able to leave a message for a person without having to actually talk with them, and in some cases, began actually preferring to leave a message. The caller would reap a new double benefit: "credit" for having called and the benefit of not having to talk at that moment. People thus began to phone when they knew they were more likely to get the answering machine than the person, thus maintaining their connection while optimizing their free time. Kids, for example, who didn't really want to speak to their parents, could call home when they knew their parents wouldn't be home, get parental praise for having touched base, and assume passive parental permission. They can tell their parents, "Well, I tried to reach you, but you weren't there so I went ahead and...." Culture changed. The answering machine transformed telephone etiquette, how the phone is used, and its intended purposes.

Other examples are devices and systems (like computers) that are designed to free up time for other things. International Business Machine's (IBM's) early computer slogan was, "Machines should work. People should think." The ways in which personal computers (PCs) have revolutionized daily life are obvious. The important point here, though, is that the time- or labor-saving promise of computers actually did not generate more leisure time, because the time or labor savings was immediately taken up with ancillary tasks that were necessary to maintain the PC. For example, while the computer made resume writing significantly more efficient and promoted customization, the so-called free time it generated was then taken up with managing software interfaces, learning new capabilities (like printing addresses instead of handwriting them), and a myriad of other skills.

According to Tenner, nature's revenge begins once the adoption of technological innovation takes root. At that moment, when most have adopted the innovation, culture is transformed and the transformation process has necessarily created a pathway to some kind of unintended negative effect (the bite-back) that works to cancel out the benefit conferred by the innovation. In other words, once culture's custom is transformed, the positive effect gained by integrating the technological innovation begins growing the seeds of adversity, which were planted during the transformation—that is, Tenner's revenge effect.

The Crystal Ball: Predicting Our Genetic Future

Despite the rapid growth of our genomic knowledge and the increasing number of available genomic-based diagnostics and therapeutics, widespread benefits are still years away, and so too is a fairly clear picture of what our "genetically transparent" lives will look like. Nonetheless, predictions abound. The expected impact on health care is immense, as previously has been discussed. Nearly all major diseases are believed to have a genetic component.[2] Ideally, improved genomic-based diagnostics and therapeutics will reduce disease risk; improve disease prevention, treatment, and management; and possibly generate cures. Some foreshadow the ability to do this by full genome sequencing at birth so that preventive measures can be applied before environmental influences increase risks. This possibility may not be as far fetched as it would seem. One company, 454 Life Sciences, is already working on speeding up testing time and reducing cost, with a goal of offering the $1,000 genome. Such a scan would offer greater knowledge at lower cost than the current battery of tests that cost between $300 and several thousand dollars. Notably, a full genome sequence is likely to be fraught with numerous ethical concerns, not the least of which is possible prenatal sequencing that could be used as a basis to coerce testing, terminate a pregnancy, or limit a child's opportunities based on the "meaning" (medical and otherwise) of the sequence (à la the movie *Gattaca*).

By 2025, some estimate that at least 50 person of all U.S. citizens will have some form of technology embedded in their bodies for the purpose of tracking and identification. Before that becomes a reality, however, some forecast that an individual's DNA profile will be imprinted on a computer chip about the size of credit card and used in place of personal identifiers.[3] The card will be privacy protected and contain specific information about your individual genetic makeup and predicted phenotype, including but not limited to probable risks of developing certain conditions, as well as key information about your metabolism based on the presence or absence of specific gene variants. The information would be intended to help you and your doctor identify effective preventative measures, select medicines that are predicted to be safe and effective for you, or even determine a customized nutrient regimen designed to promote wellness and reduce premature aging. This notion is arguably not farfetched, since the Food and Drug Administration (FDA) recently approved the first implantable chip capable of displaying an individual's health record when properly scanned.

Douglas Mulhall has an even bolder vision of our "molecular future." He envisions the convergence of molecular biological technologies, artificial intelligences, robotics, and nanotechnology to deliver more than simply the "right instruments at precisely the right time," but capabilities worthy of anointing the twenty-first century the greatest yet. He foresees what commonsense defies. Human tissue, Mulhall portends, one day will be produced from an inkjet printer. Sight will be restored by implanting microscopic computers into the eye and making connections to areas in the brain. Manufactured diamonds will run computers cheaper, faster, and cooler.[4]

Lest we be seduced into believing that biomedical innovations, or even other technologies, will predominantly define our future, they are only part of the overall forecast. For example, others predict that by 2050 a synthetic computer or machine-based intelligence will have become truly self-aware, that is, not only conscious but self-conscious. Moreover, a confluence of numerous cultural factors will contribute as well. For example, some predict that by 2100 racism will no longer be a significant phenomenon in most countries of the world.

Trends that pave the way to such ends are already under way. Mass marketing is becoming obsolete, giving way to personalization and customization, as evidenced by customized music streaming, customized media, personalized medicine, customized nutrition, individualized personal care products, customized clothing, and so on. It is curious to see that technological strategies mirror culture and vise versa. This customization craze could reflect a backlash to the depersonalization of mass marketing. For example, the 1950s and 1960s ushered in not only a leisure mentality but also an appetite for convenience that endorsed homogeneity. Mass marketing, whose origins coincided with a social policy that embraced equality and coincided with the Civil Rights movement in the United States, has spread throughout the globe, as evidenced by McDonalds, Starbucks, Gap, Kentucky Fried Chicken, and other major corporations that are found in far-reaching corners of the world. The cultural tide is changing, however, in part because current technologies enable personal isolation as well as customization.

Regardless of the veracity of these forecasts, we know that the future will be different from the present. Change is inevitable with or without human intervention. If anything in this world is certain, it is change. A corollary to this axiom is that the inevitability of change brings with it unintended and unanticipated consequences, both good and bad, and if Tenner is correct, revenge effects.

Technological Optimism: The Impetus for Our Confidence in Genetics

Americans have grown accustomed to integrating technological advances that revolutionize aspects of daily life: the Internet, cell phones, high-definition television, and robotic vacuum cleaners. Our faith in technology's ability to provide substantial benefits (over minimal detriments), that is, improve efficiencies or increase pleasure, has been steadfast since the end of the nineteenth century. The World Fairs of the 1950s and 1960s showcased the coolest, most imaginative, and seemingly effortless futures made possible by technology. This allure of technology induced our omnipotent faith in its ability to solve a broad array of problems. Such confidence implies that technological solutions come with little downside. Arguably, this was more of a social mind-set in the 1950s, but a significant confidence remains in the power of technology and medical history that has provided little evidence to the contrary.

Until the end of the twentieth century, science and technology targeted clearly defined problems. Medicine, for example, became increasing focused on identifying and solving the acute problems because technologies permitted such. Care then shifted from the whole person to the locus of the immediate or acute problem (e.g., the lungs or heart). It wasn't Mr. Jones, per se, in bed 240 on the fifth floor as much as it was the bowel obstruction in bed 240. Medical technology improved to the point it was better able to pinpoint "the problem," which in this case, meant a kink in Mr. Jones' small bowel. Improved outcomes accompanied this approach. As a result, the new instruments and devices were anointed with unprecedented power to diagnose and treat through specification and localization. As such, the technology commanded authority and trust in its ability to deliver truth, albeit a truth that was visible, acute, and localized. In other words, attention focused solely on what was amenable to measurement by these instruments. More chronic and insidious illness fell by the wayside.

The enormous success of these devices in handling the acute or sudden catastrophe has now given way to focusing on long-term subtly intertwined problems less easily defined and remedied, more gradual and global. Recent technological advancements, for example, are designed to assess the cumulative impact of the imperceptibly gradual factors that we are now realizing, are no less potent in their adverse effects. Genetics is but one example of how technology is advancing knowledge of the imperceptible. Thanks to these technological advances, we are increasingly able to understand how largely invisible and not immediately hazardous processes actually present real and potent threats, such as the x-ray machines in shoes stores that kids loved putting their feet into during the 1950s, which now are recognized as involving a pernicious, and in many cases involuntary, health risk. In medicine, this evolution has succeeded in providing more effective, less invasive, and less painful care than ever before.

Yet, despite such successes, all is not well with American medicine. Chronic diseases such as diabetes and hypertension are on the rise. Obesity in America is reaching epidemic proportions. Tuberculosis, once thought as having been eliminated, has reappeared and with resistance to traditional antituberculosis drugs. The potential for global diseases, such as human immunodeficiency virus (HIV), severe acute respiratory syndrome (SARS), mad cow disease, and avian flu, continues to become more potent with increased global travel. The spread of antibiotic resistance, now a global phenomenon, is dramatized by the appearance of penicillin-resistant syphilis in Vietnam War veterans. Germs, in other words, are outsmarting our therapeutic armamentarium. Even genetic solutions are succumbing to "bite-back" effects. Gleevec (imatinib mesylate, Novartis Pharmaceuticals), initially the darling of genomic-derived effective therapies for myelogenous leukemia, is now being beaten by tumor cells that have cunningly mutated to defy treatment. New approaches are necessary. Nonetheless, we believe that we can outsmart cancers and other diseases, but exactly how to do this eludes us, although a technologically based solution would seem a sure bet.

Unintended Technological Bite Back: New Health Risks Come with Improvement

The more you need health care, the less likely you are to receive it.
—Gwatkin Davidson, World Bank consultant

3.2 million children die each year from birth complications and neonatal infections in 42 nations.... 2.1 million children die each year from pneumonia as do from diarrhea.... The interventions that make the biggest difference are often simple ones: sterile birth kits, antibiotics for typhoid, milk, zinc, and vitamin A.... We've conquered most childhood infections but extreme reactions to everyday substances pose a new threat.
—*Newsweek*, September 22, 2003

Revenge, as Tenner conceptualizes it, is the ability of the world around us to "twist our cleverness against us." A revenge effect, according to Tenner, isn't an inherently inevitable by-product of technology alone, but rather a function of that technology embedded in a context of use: culture. Technology occurs in the context of legal statutes, state and federal regulations, standards of practice, and implementation guidelines that affect usability. The cultural context (e.g., values, beliefs, habits, and customs) further shape how technology will be adopted and its transformation of behaviors of daily life.

Consider this possible example of technological revenge. Allergies are becoming a twenty-first-century epidemic. Allergies are a heightened response of the immune system to foreign substance. For reasons not fully understood, nearly any substance can be an allergen. Genetics doesn't explain the recent rise in allergic responses, which suggests that something in the environment may be the culprit. Ironically, improved hygiene is thought to be a major contributor. The hygiene hypothesis suggests that the human immune system, which evolved in an environment teeming with hostile bacteria, parasites, and viruses, is frustrated in our modern antiseptic environment to the point it responds to previously harmless substances, like peanuts or shrimp. It's postulated that the immune system's response threshold has been lowered by lack of earlier exposure to a hostile germ-laden environment. This type of ironic phenomenon exemplifies Tenner's revenge effect. In this case, improved hygiene—though proven to have reduced, or even eliminated, some nasty lethal disease and illness—may have (if you accept the revenge theory) inadvertently led to greater and in some sense more intractable illness by crossing an unknown and imperceptible threshold of "good enough" into the "too clean" sphere.

Consider the following other examples of household improvements. Living standards have increased substantially over the past hundred years. Many household technologies have spurred improved health. Eliminating soot and smoke by replacing wood-burning stoves with cleaner forms of heating directly led to significant reductions in lung disease and consequently improved health in the populace. Central heating reduced the incidence of chilblains (skin inflammations following prolonged exposure to dampness and cold) to a rarity.[5] But home improvements designed to increase comfort can be hazardous to our health. Wall-to-wall carpeting emerged in the 1950s as a nearly universal choice for homes and offices. The technology promised warmth and cleanliness. Particularly in tighter homes and offices (in which insulation and tighter weatherproofing greatly reduce the exchange of indoor and outdoor air in efforts to keep air pollution outside), the comfort of increased warmth and humidity proved to comfortable not only for human but pests as well. Flea infestation in the early 1990s in England increased 70 percent.[6] Wall-to-wall carpets are a haven for dust mites as well as fleas. The fecal pellets of dust mites contain a potent airborne allergen that stimulates the immune system to inflame the airways. Research indicated that children in households with high levels of dust mite allergens were five times as likely as others to become asthmatic by their teens. Vacuum cleaners, long promoted for healthy living, are now thought to be a primary culprit. Not only sucking up dirt, vacuum cleaners also bounce mite pellets into the air where they stay for days and eventually settle back into the carpet; however, vacuum cleaners and mites are not the only inducers of asthma. Materials shed by fleas, dogs, cats, cockroaches, second-hand smoke, and industrial pollution are all thought to play a roll.

Will Genetics Induce Its Own Revenge?

Whether technological revenge might seethe beneath genomics applications is clearly unknown. Nonetheless, pondering potential adversity from this perspective may help to remove some of excitement and even uncritical acceptance of genomic innovations, lest we become so enamored of the promises that we are blinded to the perils.

The perverse logic to the spread of resistant genes as a direct outcome of our ability to kill infections with antibiotics seems to be a "genetic revenge" effect. Similarly, the more effective a pesticide is and the more widely and intensively farmers apply it, ironically, the faster and stronger genes mutate to confer immunity to it. DDT (dichlorodiphenyltrichloroethane), for example, was one of the first proclaimed "victors of the pesky insect." First used commercially in 1945, by 1947 DDT-resistant flies emerged. Some 10 to 15 years later, body lice were no longer destroyed by DDT treatment. Farm, orchard, and forest insects emerged that were also unharmed by DDT. The speed with which the genes of these organisms adapt is relatively amazing considering how slowly human genes have evolved over millions of years.

The emerging field of nutrigenomics offers fodder for this thought experiment. Nutrigenomics is a subfield of medical genetics and genomics that marries genomics (the structure and function of an individual's genetic makeup) and nutrition (dietary chemicals) to provide a molecular understanding of how nutrients affect gene expression at the individual level. Common dietary chemicals and nutrients can act on an individual's genome directly to alter the gene structure or function of that gene and thus gene expression. Conversely, the diet-regulating genes (and their common normal variants) play a role in the onset, progression, and severity of chronic diseases. Knowing one's genetic makeup with respect to nutrient metabolism and associated phenotypic expression can inform one about which nutrients in what amounts are needed to optimize health as well as reduce the risk of premature aging and related diseases. Recently, it has been reported that a diet low in salt in not necessarily beneficial to all hypertensives, despite having been a long-standing tenet of the treatment of hypertension. Genomic investigation will be helpful to identify this subgroup and determine why they do not benefit from low-salt diets. Ultimately, that research will permit an effective and customized treatment for this group that, in all likelihood, doesn't involve sparing the salt shaker.

Metabolic disorders provide vivid examples of how diet can mitigate disease. Phenylketonuria (PKU), discussed in prior chapters, is an inborn error of *metabolism* that is inherited as an autosomal recessive trait. PKU occurs in 1 of every 15,000 births in the United States, with incidence varying around the world. The condition results from an absent or deficient enzyme (phenylalanine hydroxylase) that metabolizes (breaks down) the essential amino acid, phenylalanine. With normal enzymatic activity, phenylalanine is converted to tyrosine (another amino acid), which is used by the body. When phenylalanine hydroxylase is either absent or deficient, phenylalanine cannot be broken down and thus accumulates in the blood, eventually becoming neuro toxic to brain tissue and resulting in mental retardation. The treatment is dietary; a carefully controlled phenylalanine-restricted intake prevents mental retardation as well as neurological, behavioral, and dermatological symptomatology.

Another dramatic example is adrenoleukodystrophy (ALD), one of several hereditary neurodegenerative leukodystrophy disorders in which the myelin, which ensheaths the axons of both the central and peripheral nervous system, is destroyed. ALD was popularized in the movie *Lorenzo's Oil*, which depicted Lorenzo Odones' case of ALD and his parent's endless search for a cure. Defying the medical establishment, the parents' experiments showed that a combination of two fats extracted from olive oil and rapeseed oil prevented the onset of disease (in presymptomatic boys) by stopping the body from producing the long-chain fatty acids whose buildup leads to the loss of myelin (demyelination). The diet, named Lorenzo's oil, has kept Lorenzo alive and he turned 29 on May 29, 2007. Lorenzo's mind remains intact thanks to the diet.

The promised benefits of nutrigenomics are great, for the promises extend to all humans. Furthermore, this rapidly evolving new field is yielding knowledge that can be readily applied by using existing nutrients. The elegance of its seemingly simple premise is that our individual genetic makeup offers crucial metabolic insights that can inform our ability to fuel our bodies with the type and quantity of nutrients that could optimize functionality at a molecular level. Hence, we could reduce and perhaps prevent a host of disease risks caused by years of metabolic inefficiencies or suboptimal functionalities by eating the right nutrients in the right amounts to optimize gene expression.

Nutrigenomics as Possibly Ripe for Revenge

This seems risk free, but is it? Well, it might be if all persons had access to adequate (optimal) nutrition and health care, access to genomic testing to identify metabolic features, and biochemical testing to chart the effects (presumably improvements) of compliance with the indicated diet. Clearly, the probability that such could be the case is exceedingly low given massive amounts of human starvation, political and economic barriers to equal access, and contentious ethical concerns. These significant hurdles may well represent formidable barriers. Regardless of how adverse the effects of these limitations might be, they are nonetheless, not revenge effects, for they are not a reaction of nature but rather a product of human control. A revenge-like effect may be brewing, however, if we look beneath the broader cultural context in which nutrigenomics is emerging.

Nutrigenomics offers the possibility of customization of one's diet at a time when individuation increasingly describes the retail marketplace including health care. Genetics—referred to as the science of difference or discrimination, in that it characterizes defining properties that distinguish species and individuals from one another—offers an illustration of how technology and culture can mirror one another. Viewed in this way, the parallel emergence of genetics and personalization or customization is rather compelling. Personal autonomy was given a big boost by technology, as evidenced by the emergence of PCs around the time that recombinant DNA technologies began to emerge. Recent advances in genetics further spur this tidal wave by evincing the exact how and why of difference. That customization, with its emphasis on the importance of defining differences (genetic and not), is ushering in a cultural transformation, and with it, a facet of technological revenge may be emerging.

The exquisite and speedy genetic resistance of pests would seem to qualify as a genetic revenge effect to human efforts (or hubris) to control discrete and specific aspects of the ecosystem. As health care moves toward increasing personalization, one wonders whether this paradigm shift is likely to induce analogous untoward effects. Feeding our genes with supplements that destroy the optimal homeostasis nature has achieved over centuries may provide a potent additional example. The American diet, rich in saturated fatty acids and cholesterol, is believed to negatively contribute to the high levels of cardiovascular disease, diabetes, and hypertension. In efforts to reverse these effects, Americans now are being advised to decrease saturated fats (found in meat, dairy products, and some tropical oils) and trans-fatty acids (made during the hydrogenation of vegetable oil) and instead eat polyunsaturated fats, like omega-6 and omega-3 acids. The most healthful of the omega-6 fatty acids is linoleic acid, which converts gamma linoleic acid (GLA) to prostaglandins, hormone-like molecules that help regulate inflammation and blood pressure as well as heart, gastrointestinal, and kidney functions. Good dietary sources of omega-6 fatty acids include cereals, eggs, poultry, most vegetable oils, and whole grain breads, although many take omega-6 in the form of supplements. Omega-3 fatty acids are found in canola and olive oils, walnuts, green leafy vegetables, flaxseed, walnuts, and coldwater fish (salmon are well known for their health benefits). Omega-3s appear to improve insulin sensitivity in noninsulin dependent (or type 2) diabetes, ease menstrual pain, play a role in keeping cholesterol levels low, stabilize irregular heartbeat (arrhythmia), reduce blood pressure, improve rheumatoid arthritis, lupus, Reynaud, and other autoimmune diseases, aid in cancer and stroke prevention, reduce inflammation, and even relieve depression and symptoms of other mental health problems. The body cannot make its own omega-3 and omega-6 fatty acids, so health requires ingesting them. While many are convinced of the health benefits of polyunsaturated fats, omega-6 fatty acids are being added to the diet in greater quantities than omega-3 fatty acids, and the disparity is disrupting the body's inherent optimal balance, thus impairing the expected health benefit of adding omega-6s. Efforts to combat a problem, cause a bite-back effect because the natural balance is destroyed.

Arguably, more potent revenge effects could occur on a larger public health scale. Do individuals have the right to not avail themselves of new genetic knowledge and its indications for optimum health? Are we heading into a world where individuals do not have a right to be sick or a right to be disabled as they've chosen to not adhere to genetically based prevention? One can conceive of other public health measures that have been in place for years for the benefit of the community starting to fail, if the community loses some of its communal biological profiles from increased personalized and nutrigenomics, whose use may have adverse biofeedback consequences. For instance, as more people consume a customized diet, untoward effects could occur on something as basic as blood types, making the regular blood typing suboptimal for determining the safety of transfusion because of greater microvariation that could be incompatible with another's biochemistry or physiology. The public health principle behind vaccination is that vaccines target disease (whose symptomatology is functionally the same in large groups of individuals). Were the applications of genomic advances to customize and optimize individual health at a micro level with profoundly positive effects, it is *possible* that the herd for a targeted vaccine would be too small to make an impact. In other words, using genomics to optimize individual health may result in impeding certain communal goals of public health measures as well as in clearly pitting individual right's against the norms of the community.

Finally, threshold effects could induce technological bite-back effects. It is possible that we will see the therapeutic benefits associated with genetic advances, such as more personalized nutrition or more personalized medicine, reach a threshold or plateau effect, after which the therapeutic benefit is no longer realized. For example, people starting an analgesic may get therapeutic relief from an initial low dose, but continued use induces tolerance such that a higher dose is required for a therapeutic response with the possibility of novel adverse effects. Similarly, the arguable increased, even over, use of antibiotics is thought by some to be responsible for current antibiotic-resistant bacteria, resulting in fewer effective antibiotics. The development and use of new antibiotics may well be accompanied by new adverse effects. Physicians sensitive to this dilemma are increasingly reluctant to prescribe antibiotics. It seems far from obvious that personalized medicine or nutrition could be immune to such threshold effects. Assuming the possibility of such threshold effects raises the larger question of whether the solution will achieve the desired benefit. In other words, our assumption that genetic advances will continue to improve health may come with qualifications. Genetic technologies might create some bite-back effects.

Normative Issues

A core ethical issue arising from such conceptualized genomics bite-back effects is an implicit paradigm shift from a commitment to the collective (or social) to the supremacy of the individual. To optimize individual health only to conceivably compromise public health measures may be rationale on some levels but ethically repugnant on others. Gone could be a certain motivation toward altruism, particularly if effective strategies to achieve such ends became distinctly restricted, if not impossible. Blood typing and vaccination are but two ways in which individuals *are* able to benefit others, particularly at little cost to themselves. To compromise such avenues could potentially usher in an entirely new foundation of ethicality, and one that could sadly break roots with fundamental moral principles, such as duties to others.

Nonetheless, at the heart of the problem is an implicit notion that we are somehow "entitled" to progress. Indignation about failures of technology to achieve genuine progress naturally follows from this presumed entitlement. This tacit notion of entitlement to progress would seem tautological, except for the fact that it's rarely acknowledged, let alone justified.

From the time of ancient Greece, we've believed that knowledge is a good thing and that acquiring new knowledge as well as applying that knowledge to improve the world is also a "good thing." Furthermore, striving to know more and use that knowledge in benevolent ways that benefit the maximum number of people is generally considered a good thing. Typically, at issue is not whether we should create more knowledge, but rather what we ought to do with it once we have it. The question of who is free to obtain genomic knowledge, who has the right to refuse to obtain that knowledge, and who has the right to enforce that knowledge to protect self and others from harm are questions that have been analyzed within the previous topical chapters. Although the capability and promise of genomics continues to garner favor, it is too early, arguably, to determine whether in the end, genomic knowledge will have been a "good thing." In addition to anticipating and proactively guarding against the potential for adverse consequences, it may behoove us to consider Tenner's revenge theory and with it the warning that yesterday's miracle solution potentially could become today's nemesis.

Notes

1. Tenner, Edward, *Why Things Bite Back: Technology and the Revenge of Unintended Consequences* (New York: Vintage Books, 1997).

2. Collins, Francis, director of the National Human Genome Research Institute: "Virtually all diseases, except maybe trauma, have a genetic component," see science-education.nih.gov/nihHTML/ose/snapshots/multimedia/ritn/vhl/disease.htm.

3. www.ourmolecularfuture.com; www.foresight.org/nanoecology/Mulhall.html www.dnatoday.com/packages.html; www.rsc.org/chemistryworld/restricted/2004/June/matchmaker.asp; and www.marl.mb.ca/f.php?PUBLICATIONS/talk5.

4. Maney, Kevin, "Man-made Diamonds Sparkle with Potential," *USA Today*, October 6, 2005, http://www.usatoday.com/tech/news/techinnovations/2005-10-06-man-made-diamonds_x.htm; www.longbets.org/predictions; Mulhall, Douglas, *Our Molecular Future: How Nanotechnology, Robotics, Genetics, and Artificial Intelligence Will Transform Our World* (Amherst, NY: Prometheus Books, 2002); see also www.ourmolecularfuture.com.

5. Schwartz, Ruth, *More Work for Mother: Ironies of Technology from the Open Hearth to the Microwave* (New York: Basic Books, 1983).

6. Tenner, Edward, *Why Things Bite Back: Technology and the Revenge of Unintended Consequences* (New York: Vintage Books, 1997), 131, 140.

APPENDIX A

Ethical Decision-Making Exercises

Chapter 3 Exercise

Simulate the experience of deciding whether to undergo genetic testing. Pay particular note of personal considerations, specifically those related to your genetic privacy. Consider the possible impact on yourself and on your relatives, close friends, and potential spouse.

Learning objective: Simulate decision making about whether or not to take a genetic test and grapple with ethical issues that arise.

Have a classmate gather five sheets of paper. Without your seeing the specifics, ask the person to write a genetic diagnosis at the top of three sheets and seal each diagnosis in a separate envelope. Example diagnoses can be found in chapter 3 or can be found in a basic human genetics or medical genetics textbook. Then ask the person to write two genetic risk-based diagnoses on the remaining sheets (for example, an inherited 86 percent probability of cardiac failure at age 57 and obligate carrier of cystic fibrosis). Shuffle the sealed envelopes. Take one of the envelopes at random and put it in front of you. Access the Internet and go to the http://genetests.org Web site.

Learn about the current genetic tests being researched and available at a clinic. Using the two diagnoses, now answer the following questions:

1. Do you want to know your genetic profile? Yes ___ No ___ Don't know ___

2. Do you believe that you ought to know your genetic profile before deciding whether to marry? Yes ___ No ___ Don't know ___
3. Do you believe you ought to know your genetic profile to decide what life occupation you want? Yes ___ No ___ Don't know ___
4. Do you want to know the genotype of your fiancé before you commit to marrying that person? Yes ___ No ___ Don't know ___
5. Do you think you should know your genetic profile to decide whether or not to have children? Yes ___ No ___ Don't know ___
6. Do you think you should tell your relatives if you find out? Yes ___ No ___ Don't know ___

Make your decisions. If your answer to question 1 is yes, then open the envelope and answer the following:

1. Are you happy with your decision? Yes ___ No ___ Don't know ___

2. Repeat questions 2–6 above. Are your answers the same after knowing your genetic profile?

Repeat this exercise until you've examined all three envelopes.

Chapter 4 Exercises

Exercise 1

Assume that a new genetic test for predisposition to violent behavior is commercially available and that a positive test result indicates a probability that the individual will commit a violent act. Your state has just passed a law requiring that all residents age 18 and older must be tested. Individuals may be tested at their primary care physician's office. Those without health insurance may be tested at mobile laboratories in designated areas in addition to the State Laboratory. The State Lab will conduct the testing and retain all DNA for possible future behavioral and forensic testing; it is also charged with securing the privacy of the data bank according to existing federal and state laws. The law further mandates that individuals who test positive will be notified and required to register with local police stations as well as update records about place of residence and employment status. Databases tracking marriage, births, and deaths will be merged with the police DNA bank to maintain updated records of registrant's marriage and parental status.

1. Identify the ethical arguments in favor and opposed to this require-ment using the theories and principles discussed in chapter 2.
2. Discuss whether the ethical analysis itself would change, and if so, how it would change depending on the probability of criminal activity. That is, assume a positive test result confers a 10 percent probability that the individual will commit a violent act. Then assume a positive test result confers a 90 percent probability that the individual will commit a violent act. Compare and contrast the ethical justifications for policies based on the 10 percent probability versus the 90 percent probability.

Exercise 2

The goals of this exercise are to understand a few of the fallacies of the eugenicists and to experience some problems in defining behavior, including the following: researcher bias, predictive results, and how pre-dictive results could be used for social ends.

Collect Evidence

1. Think of all the people you know well.
2. Of this group, has anyone ever had an explosive outburst of any kind? An explosive outburst could be defined as any of the follow-ing: a temper tantrum after losing a game, yelling, slamming a door, stealing, running away, or shouting when not getting what they wanted. If no one has exhibited this type of behavior, can you be sure that in the absence of your observation they never have exhibited such behavior? You may not have actually witnessed the event, but know from hearsay. Write the name of each person and draw a box around the name.
3. Think about their relatives (siblings, progeny, parents, grandpar-ents, etc.). How many of them have exhibited such behavior, even if less often or less extreme? This could include the following: yelling at a another driver on the highway, shouting at a police officer, being angry and saying harsh words to a store clerk, or expressing outrage when reading something in the newspaper.

Establish a Probable Cause

1. What do you believe is causing these outbursts?
2. As your teacher, I believe that the ingestion of acid (i.e., the acid in orange juice) in the morning is a cause and that regular

ingestion may cause a build up of acid in the body, resulting in the permanent loss of voluntary control over anger.

3. Enter an X in the box if the person drinks juice in the morning every day or regularly.

Analyze the Evidence

1. Is there evidence of violent behavior?
2. Does anything seem silly here and if so, then why? What's right or wrong with this line of reasoning?
3. Was anything proved in this exercise? Why or why not?

What might be wrong here? How is criminality defined? Is your sample representative of the general population? Is your sample self-selected? Did you include only examples that illustrate your point, and thus the data are taken out of context of actual population frequency? If your observable trait occurs frequently, does frequency prove that the cause is genetic? (E.g., if there is a high incidence of truancy on Cape Cod, does the incidence prove that the cause is genetic?) Have you considered alternative explanations, such as nongenetic factors? Davenport presumes biological determinism, and doesn't consider environmental influences, for example, that imbecility might have resulted from a childhood head injury or lead exposure.

Is there evidence of a precise pattern of inheritance (e.g., autosomal dominant or recessive, or polygenetic)? Define these terms. Is there identification or acknowledgment of any confounding factors? What about the overall merits and applicability of the work? Who paid for the research? Whose interests are at stake? Are the results likely to benefit or be profitable for the researchers? What is the sample size? Is it large enough to extrapolate to the population at large and be meaningful? Were the results published in a respected publication? Are the limitations of your research or any negative results reported? All of these questions are important to ask to ferret out the validity of purported claims.

Has genotype truly been inferred from phenotype? Isn't it possible that identical phenotypes refer to distinctly different genotypes? For example, mental retardation might result from congenital anomalies or oxygen deprivation at birth. The cause may differ even though the outcome (phenotype) is the same. Furthermore, have you proved that the acid causes criminality? Indeed, you may have shown an association, but an association is not a cause of the behavior. To show a cause, you would have to test other antecedents to see whether they also produce criminality, like eating chocolate. Lastly, can you generalize that from this data set that anyone drinking acid in the morning is likely to become aggressive or violent? Why or why not? Taken together, is your reasoning circular? Do the arguments beg the question? Do the arguments indicate the fallacy of appealing to an authority, as in someone in authority says so and therefore it must be true?

Chapter 5 Exercises

Exercise 1

1. Pick a diagnosis and research its standard treatment(s), and then write them on a piece of paper. On separate pieces of paper, name different pharmacogenetic test results with half of them being positive and the other half being negative. (This means that 50 percent will show that the treatment isn't safe [list the reasons], but it will be safe and effective for the other 50 percent.) Insert each choice into a separate envelope, seal it, and mix them up into a single pile.
2. Pick one envelope and make a choice of whether open it. (a) If you have opened the envelope, what are your thoughts now that you know the test result? If you test positive, project how you would feel if this test in the future indicated that you have the single nucleotide polymorphism that is associated with an elevated risk of cancer (breast cancer in women, prostate cancer in men). What is your reaction to learning genomic information that you did not originally consented to knowing? (b) If you did not choose to know your diagnosis, ask yourself what would happen if you began having serious symptoms (define them). Would you regret not taking the test?
3. Describe the scenario with your primary care physician. Would there be any emotional pressure to bear to choose one way or the other? Consider your thoughts if in the future you were diagnosed with cancer (breast cancer in women, prostate cancer in men), and the disease could have been prevented if you had known about this predisposition in earlier years.

Exercise 2

Your doctor is prescribing antidepressants. Because you've never taken them and don't know how your body will react, you are advised to take a quick pharmacogenetic test to determine which of several available drugs are likely to be safe and effective for you. The test itself is easy.

All you need to do is swab the inside of your cheek with a swab resembling an elongated Q-tip. The results are available in one week.

Take out a blank piece of paper and pen to answer the following questions:

1. Do you want to take the pharmacogenetic test? Yes or No? Why or why not?
2. Do you have any consent or privacy concerns? Why or why not?
3. The test will tell you only about 7 or 14 available drugs. Do you still want to take the test?

Assuming you've undergone the test, your results indicate that six of the seven drugs the test analyzes are not likely to be safe or effective. The seventh, however, is likely to be both safe and effective. The seventh drug is a relatively new and very costly drug. You would have to pay a copay of $150 for a monthly dose of this medication (the total cost is $639). As a college student without much money, and without parents who will pick up the tab, what drug do you want to try? Realize that the pharmacogenetic results are probabilistic.

Assume you decide to try one of the cheap but less likely to be safe and effective medications, because even though unlikely, it may work. Your doctor, however, refuses to prescribe any of these six drugs for fear of a malpractice suit if you experience an adverse event. What ought you to do? Why?

You accept a summer job with a self-insured employer, which requires a physical exam as a condition of providing health insurance. Part of the exam requires taking a pharmacogenetic test for commonly prescribed drugs. The employer requires the test because it wants to avoid paying for drugs that are not likely to work. Your job is not contingent on taking the pharmacogenetic test, but you will not be given health insurance if you refuse the test. What ought you to do and why?

Chapter 6 Exercise

Divide the class into fourths. Group 1 will be biomarker researchers. Group 2 are research subjects. Group 3 are family members to Group 2. Group 4 are internal review board (IRB) members charged with approving the research protocol and ensuring institutional compliance with policies and practices governing strict privacy and security of stored samples.

The Study

Background. Fatty acids represent a major storage form of energy. When caloric intake exceeds energy demand and glycogen stores are saturated, excess dietary carbohydrate is converted to fatty acid and stored in adipose tissue as triglyceride, together with excess dietary fat. When energy demand cannot be met by the utilization of stored glycogen, fat is mobilized from adipose and transported to tissues such as muscle, where it is taken up and oxidized for energy production. Thus, the rate of de novo fatty acid synthesis is rapid in well-fed people and slow in fasted people.

Statistically, African-American women are at greater risk of heart disease than Caucasian women. Significant differences between African-American and Causcasian men exist with respect to heart disease risk and successful outcomes. In general, black men experience earlier onset of disease, more severe disease, higher rates of complications, and more limited access to medical care than white men. Native Americans are a lower risk for heart disease based on their fatty acid metabolism.

The Research. The goal of this study is to identify the protein (ApoE) and its different allele's role in fatty acid metabolism. ApoE assists in the processing of lipoproteins and fatty acids. We discovered that the three different alleles (ApoE2, ApoE3, ApoE4) have different effects on the uptake of free fatty acids (a type of fat) by cells, which might contribute to their differential effects on the development of atherosclerotic disease. Free fatty acid uptake is increased with ApoE2 and decreased with ApoE4 as compared with the uptake in the presence of ApoE3, the most common form of ApoE. The studies in this proposal are designed to determine the reason for these effects on fat metabolism and to examine the effects of these differences on the metabolism of fat in experimental animals. The results of these studies will increase our understanding of the role of ApoE in fat metabolism and in the prevention of premature atherosclerotic disease.

Group 1: Your Research

Purpose. You are undertaking a cross-sectional study to clarify associations of lifestyle factors (habitual exercise, alcohol intake, and smoking habit) and plasma fatty acid (FA) concentrations. You are using adiopose tissue FA concentration as a biomarker for atherosclerosis. You are also doing genotyping for ApoE, which has been found to prevent heart disease. In humans, ApoE exists in three different forms, called ApoE2,

ApoE3, and ApoE4, which vary by single amino acid changes in their structure. These forms of ApoE differ in their effects on lipoprotein metabolism and atherosclerosis. The mechanisms that lead to these differences are not completely understood. Your research will investigate both correlations of the biomarker and clinical effects as well as correlations with the different forms of ApoE, as the different forms have different effects on the uptake of free fatty acids.

Method. You will collect diet records, lifestyle information and blood samples from men and women who are self-identified Caucasian, African American, Asian, or Native American. You will analyze dietary FA intake and analyze plasma FA concentrations as well as store samples for a large population study of fatty acid concentration and ApoE alleles. The blood will be tested for the biomarker and the genotype (different ApoE alleles). You know that Europe is about to embark on a similar study and that the National Institutes of Health (NIH) is interested in eventually combining data from your study with that of studies in Europe, so you want to store samples for future testing.

Tasks

Group 1. Write an informed consent form for potential subjects, which includes consenting to participation and storage of samples. Participants can participate in the study and refuse to have their samples stored. Similarly, they can agree to have their samples stored but not be recontacted. It is their choice, and permit them to indicate such on the consent form (you will need to refer to their decision later). You can decide whether you will have stored data identified, deidentified, or anonymized.

Group 4 (IRB). Review the informed consent for its handling not only of the research to be conducted but for its overall ethical acceptability of handling stored samples. Decide whether to approve the research or not or approve it conditionally, meaning that some terms must be changed to be more ethically acceptable—for example, terms of recontacting subjects.

Group 1. Meet with Group 2 and assess their enrollment criteria. See whether their self-reported ethnicity fits your study criteria. For those who fit the criteria, give them the consent form and try to enroll them. You need a minimum of five subjects to run the study.

The study is under way.

Give Group 2 a short questionnaire about diet and life style, including questions about nutrient intake, smoking and drinking habits, and exercise.

Group 3. All individuals in Group 3 are relatives of the people in Group 2. You've been told that your relative in Group 2 is enrolled in a biomarker research study. You know that you are at increased risk for arthersclerosis and that you have an immediate family member (sibling or grandparent) who is in the early stages of Alzheimer disease.

Group 1. Meet with Group 3 and attempt to enroll them in the study—give them the consent form and explain that the stored samples will be anonymized.

1. How many enrolled? Why do or don't they enroll?
2. Do you want to adjust the study midstream to adjust to these results (expand or restrict the study, etc.)? What problems arise in trying to change the study? Are there issues that you've learned from the data collected that justify a change?

Step A

Group 1. Your results are in. Group 2 and 3 are to be divided into three separate groups: A, B, and C.

Instructor. Prepare sealed envelopes for groups A, B, and C. Envelope A contains the results that you have ApoE2 and your free fatty acid uptake is increased due to the activity of the ApoE protein. Envelope B indicates that you have ApoE3, the most common form of ApoE, which is associated with average risk of arthersclerosis and is associated with promoting nerve growth after injury. Envelope C reveals that you have ApoE4, which indicates you are at high risk of heart disease. The good news though is that research indicates for this genotype, ApoE4, your risk of heart disease can be significantly lowered by increasing your intake of omega 3 fatty acids.

Step B

Instructor. Hand out envelopes A, B, and C to each group: one-third of each group gets envelope A, one-third gets envelope B, and one-third gets envelope C.

Step C

First, Groups 1, 2, and 3 open the envelopes. Each group should discuss the different results internally. Second, form new groups according to the results: Group A, Group B, and Group C. Each of these groups should discuss individuals' reactions to the results, and the similarities and differences in their perceptions of their abilities, opportunities, future, and so on.

You are many years down the road and new information is available from other studies. Homozygosity for ApoE4 is now known to be a risk factor for Alzheimer disease. There is no known treatment for mitigating risk. You nonetheless feel that some of the research subjects who tested positive for ApoE4 would want to know about their disease risk.

1. What do you do and why?
2. For those who wanted to be recontacted, would you still want to be recontacted if the news was bad and at present there was nothing you could do about it?
3. For the ApoE4 group, do you want to be recontacted by the researcher to know about your risk? Why or why not? What does it feel like to know that you have information (ApoE4 and Alzheimer are big media stories) you did not consent to originally?
4. What ethical issues arose in this case? What competing rights and interests emerged? Did those rights and interests change over the course of the research? Did you resolve issues adequately? Why or why not?

Chapter 7 Exercises

Exercise 1

Test your knowledge and beliefs about GM foods by answering the following questions.

1. People in the United States already eat a significant amount of genetically modified food. What GM food have you eaten during the past week?
2. You are standing in the vegetable section of your favorite grocery store. You want to buy some tomatoes for a dish you're preparing tonight. In front of you is a wide assortment:

   • Charlie's Mortgage Lifter Tomatoes
   • Kentucky Beef Steak

- Yellow Cherry
- San Marzano
- Pachino
- Hydroponic
- Green Grape
- Green Zebra
- Brandywine
- Amish Paste
- Cherokee Purple
- White Wonder
- Omar's Lebanese
- Costoluto Genovese (heirloom from Italy)
- Druzba (heirloom from Bulgaria)
- Giant Oxheart

3. Which one(s) do you want to buy and how do you decide?

Step 1: How much of the food you eat is genetically modified?
Step 2: If your food was labeled, would you choose GM? Yes ___
No ___ Don't know ___
Step 3: Define GM food.

4. Regarding the list of tomatoes above:

Step 1: List three criteria for your choices.
Step 2: Order those criteria from most to least important.
Step 3: Pick two types.
Step 4: List up to five factors that you didn't consider but that might be involved in your decision-making process. Some factors you might have considered are price, ripeness, color, bruise spots, shape, recommendations, and recipe requirements.
Step 5: Did you consider the nutritional content?

- Did you consider whether some tomatoes contain more lycopene and other nutrients than others?
- Did you consider whether the tomato was grown organically? Do you know what the term "organically grown" means?
- Did you consider whether it was grown hydroponically (grown without soil using chemically treated water)?

- Did you consider whether it was labeled?
- Did you consider the contents of the labeling?
- Did you consider whether the tomato was genetically modified?
- Did you consider whether the tomato was grown where the grower used pesticides (or even how many pesticides)?
- Did you consider whether the tomato has disease resistance?

Step 6: Are any of the above varieties hybrids?
Step 7: Do you know whether you've ever eaten any GM food?
Step 8: Do you care whether you eat GM foods?
Step 9: Is it acceptable for anyone to eat GM foods? Should starving people in the developing world eat GM foods if the foods have a longer shelf life and they're more expensive?

Exercise 2

Case 1. The state of Alabama's budget shows a projected deficit of $2 billion dollars growing to $6 billion dollars by the end of the decade. One effort to reduce the deficit is to reduce health care spending by public health campaigns to adopt customized nutritional programs, that is, feeding your genes what they need to improve health and lower disease risks. The state's Medicaid and Welfare programs, both of which are grossly over budget, have mandated that enrollees undergo nutrigenomic testing and adhere to individualized nutritional reports to promote improved health outcomes and lower doctor visits and thus reduced health care expenditures. Failure to comply with programmatic dictates will result in benefit loss.

1. Make an ethical argument defending the state's actions. On what basis is the state morally justified in adopting and enforcing such policies? Argue the case that involuntary testing is morally sanctioned for greater individual and societal benefit. Ought there to be exceptions to programmatic mandates and if so on what grounds?
2. Make an ethical argument opposing the state's actions. On what basis are the state's actions morally indefensible? Argue the case that involuntary testing is immoral, even for greater individual and societal benefit.
3. Discuss the relative strengths and weaknesses of cases favoring and opposing the state's actions.

Case 2. Monsanto has just launched the sale of collard greens that are genetically modified to reduce inflammation processes associated with cardiovascular risk in African Americans. Labeling is not required so consumers will not be able to easily determine whether or not their collard greens are genetically modified, but grocers are advertising that they are selling the new variety. Clinical research demonstrates the safety and efficacy of this new food in African Americans having a particular genotype. In Alabama, Medicaid and Medicare are offering cash rewards to African Americans who purchase these collard greens, based on research findings.

1. Are Monsanto's actions ethical, even if legal? On what grounds?
2. Are Medicaid and Medicare's actions ethical? On what grounds? Is it ethically defensible to market to an ethnic group which may or may not have the genotype that could benefit from the new food? Why or why not?

Chapter 8 Exercise

Step 1: Have every student take out a sheet of paper and write their views of cloning on the paper. Pro or con, they must choose one or the other. (a) Ask students if they want to qualify their response by adding certain exceptions and permit them to add the exceptions. (b) Have each person write their name at the top and pass forward to the teacher.

Step 2: Simulate decision making. Each student will be confronted with an unexpected situation. Prepare three sets of envelopes: A, B, and C. Each envelope should contain a different scenarios. Put scenario A in envelope A and seal it. Divide the class in thirds and give A, B, and C, respectively, to a third of the class. In addition, give every student a blank sheet of paper to answer the questions.

Step 3: Each person in each group should open their envelope and read their scenario. Each person should take a few minutes and answer the questions based on their situation (scenario).

Step 4: Appoint a leader for each group A, B, and C. Each leader should engage in a group discussion to find out how individuals in the group decided. Did everyone decide to do the same thing? To what extent did opinions and decisions differ? What were the points of overlap and difference? Each group leader then comes to the front of the class to summarize their group's experiences.

Step 5: Connect fantasy with reality. Here the instructor should lead ask students how their thought experiment (simulated decision making in the particular scenario) cohered or didn't cohere with their prior stated views and justifications for their positions on cloning. Note and discuss what discrepancies arose; discuss differences between views before the scenario exercise and after the exercise. It is highly likely that at least some students will have changed their views when confronted with specific situations. The instructor needs to probe for these discrepancies and the reasons underlying them. Ought personalizing aspects of decision making and position taking carry significant moral weight? Why or why not? What has been learned about discussing the issues in abstract versus a highly personal and arguably more directly relevant situation? What does this difference mean for determining what's ethical?

Scenario A

It is year 2020 and you and your partner are planning a family. You have already gone through several rounds of the most current reproductive options. Your partner has cancer (if male has a history of testicular; if female has ovarian), which affected him/her in his/her 20s; however, before receiving chemotherapy and radiation therapy, sperm/eggs were collected and frozen for future use. In the intervening years, your own health has deteriorated such that you have developed premature ovarian/testicular failure. There still is a chance for you and your partner to have a child that is biologically related to you, but it involves cloning. While there are other options, such as retrieving oocytes/sperm and undergoing in vitro fertilization with subsequent implantation of fertilized eggs into a surrogate mother, these are not possible for various practical and medical reasons, such as the fact that the bank was destroyed by fire and your tissues (sperm and eggs) have yet to be found. Given the current health of you and your partner, these other options are no longer viable, leaving cloning as the only possible strategy. Adoption is not possible because you lack the funds required and no adoption agency is likely to permit either of you to adopt because of your health status.

1. Should you pursue this option? From your reading, what were the major issues confronting reproductive cloning in the beginning of this century?
2. What should you do if you and your partner strongly disagree?
3. What ethical challenges might you face bringing a cloned child into the world?

4. Before this exercise did you believe that cloning was an appropriate thing to do? Why or why not? Did this exercise change your views? Why or why not?

Scenario B

You are a conservationist zookeeper in a tropical rainforest in Upper Guinea rainforest region, home to the last two West African Red Colobus Monkeys, previously thought to have perished at the hands of hunters. Nonetheless the World Conservation Union classified the monkeys four years ago. It is feared that other large primates in the region are becoming, if not already, extinct, primarily as a result of human population growth, hunting, and economic development. The Union and your zoo were approached by a nonprofit to save the species organization that uses demonstrated cloning techniques to repopulate species that are slated for extinction. The organization has approached you and seeks your authorization to clone the Red Colobus Monkey at no cost to the zoo or the conservation land it inhabits.

1. Should you authorize the cloning of the monkeys? Why or why not?
2. Are there advantages to preserving the species through cloning?
3. Would it be ethically acceptable to permit the monkey to become extinct for ever? Why or why not?
4. The governments of Ghana and the Ivory Coast are engaged in final stage negotiations to sell the vast majority of the tropical forest timber and botanicals for drug and health promotion supplement development. The overwhelming health needs of the vast majority of people in these countries (basic needs of adequate food, clean water, vaccinations, and health care) have been in the hands of these countries' governments who promise to use the money they get for tropical forest for these needs. However, government corruption suggests the possibility that the promise will go unfulfilled. You decide that the red monkey and other dying species in the forest must be preserved to maintain biodiversity and a balanced ecosystem, and to preserve the monkey for future generations to come. You intend to argue this point to an international tribunal. What are your arguments favoring cloning and the preservation of species in the forest (opposing the government's interest in economic development)? Who are the stakeholders in this decision? What are the competing interests? Is it possible to

fairly balance them, or must one choose between cloning, animal preservation, and economic development?

5. Before this exercise were you in favor or opposed to cloning? Why or why not? Has this experience changed your views? Why or why not?

Scenario C

You and your partner have a 3-year-old who lies, possibly dying, in a coma. You daughter (Lisa) was kidnapped 2 months ago and found barely alive. You and your partner desperately wanted a child and had enormous difficulty conceiving. Your daughter was born after 15 years of trying every conceivable strategy. At Lisa's birth you banked cord blood in case of a possible future disaster. You and your partner realize that Lisa is more likely to die than recover. You've been advised that after her death you could (legally and medically) clone her using the banked cord blood.

1. What do you want to do, clone or not clone, why or why not?
2. You've decided to go ahead with the cloning. Genotyping of the stored blood reveals considerable increased risk for breast or ovarian cancer and osteoporosis. During the cloning process, genes involved in conferring this risk can be silenced without any adverse consequences to the new child's health or wellbeing, essentially eliminating these risks. Ought you to authorize gene manipulation to lower health risks? Why or why not?
3. Before this exercise were you in favor or opposed to cloning? Why or why not? Did this exercise change your views? Why or why not?

Techniques Used in Genetic Engineering and Genetic Modification

Genetic engineering or modification refers to a set of specific techniques that alter the genetic makeup of a living organism, such as a plant, animal, or bacteria. These techniques introduce, enhance, or delete a particular characteristic in an organism by altering the organism's DNA [deoxyribonucleic acid] (genetic makeup). The difference between genetic engineering techniques and those of older conventional cross-breeding methods is that genetic engineering uses much more precise information, specifically exactly which genes and sections of DNA to target and alter to ensure a particular outcome. Older techniques involve interbreeding of species (typically based on phenotypic properties) with the hope of obtaining a near-perfect result. These genetic engineering techniques have been used on organisms to create medicines, vaccines, fibers, animal feeds, foods, and ingredients in foods.

Genetic modification techniques are thought to take much of the guesswork out of the breeding process and that is one of several reasons why they are supported. A wide range of techniques are used to modify an organism's genetic makeup. Some commonly used techniques are as follows:

Bacterial Carriers

Agro-bacterium easily delivers DNA and is used in the following ways: (1) it is prepped in a solution to make its cell walls more porous;

(2) it receives the selected gene (it is inserted into the solution); (3) it is heated to permit the plasmid to enter the bacterium; and then (4) it is tested to allow the bacteria to infect the target plant and deliver the new gene.

Biolistics

The selected DNA is attached to microscopic particles of gold or the metal tungsten by using a burst of gas under pressure that shoots the DNA into the target cells of the organism.

Calcium Phosphate Precipitation

The selected DNA is exposed to calcium phosphate that results in the creation of tiny granules that transport the DNA. The targeted cells respond by surrounding and ingesting (phagocytosis) the granules, thus ingesting the new DNA.

Electroporation

Target cells are immersed in a special solution containing the selected DNA. A short but intense electric shock is then passed through the solution, causing small tears in the cell walls. These tears permit the new genetic material to infuse itself into the nuclei. After this is complete, the target cells are placed in a different solution that prompts the repairing of the torn walls, thereby sealing the new DNA inside.

Gene Silencing

The gene responsible for the organism's undesirable trait is identified and rendered nonfunctional by attaching a second copy of the gene the wrong way around. This technique is used to prevent plants (like peanuts and wheat) from producing proteins commonly responsible for human allergies.

Gene Splicing

The desired gene is clipped using restriction enzymes taken from bacteria. These enzymes form part of the bacterium's "immune system," so if the bacterium is invaded by another organism (such as a virus), the restriction enzymes attack and destroy the organism's DNA by cutting it into precise sections. The cut ends of the DNA are "sticky" in some cases, which mean that they can be "pasted" directly onto the target organism's prepared DNA. A subsequent enzyme is needed to fuse the new gene into the chromosome. Alternatively, instead of "pasting," the new gene may be inserted into a bacterium DNA molecule (plasmid), which then invades the target cell and delivers the gene. Not all foreign DNA needs to be integrated into the host's DNA for it to be functional. In this situation, the novel DNA does not need to integrate with the host's DNA.

Lipofection

Small bubbles of fat called liposomes are used to carry selected DNA. The target cells and the liposomes are placed into a special solution wherein the liposomes merge into the cells, thereby delivering the new DNA.

Microinjection

The selected DNA is injected into a fertilized ovum (female egg cell) through a glass capillary tube. The genetically modified egg is then transplanted into a prepared uterus and allowed to grow to term. This method ensures that almost every cell in the developing organism's body contains the new DNA.

Viral Carriers

A virus with a DNA genome can be used to deliver novel genes to cells without damaging or killing the cell, after its noxious genes have been deleted. Infection of the donor cells effectively delivers the DNA to the cell and all subsequent cells. A limitation to this technique is that not all cells can be infected if the rate of division is very low, for example, retroviral vectors in the central nervous system where the neurons are postmitotic.

APPENDIX C

Updating the Nuremberg Code

The Nuremberg Code has been updated by the Declaration of Helsinki. Below are the basic principles governing human experimentation (available online at http://www.nihtraining.com/ohsrsite/guidelines/helsinki. html).

World Medical Association Declaration of Helsinki

Ethical Principles for Medical Research Involving Human Subjects Adopted by the 18th World Medical Association (WMA) General Assembly Helsinki, Finland, June 1964, and amended by the following:

- 29th WMA General Assembly, Tokyo, Japan, October 1975
- 35th WMA General Assembly, Venice, Italy, October 1983
- 41st WMA General Assembly, Hong Kong, September 1989
- 48th WMA General Assembly, Somerset West, Republic of South Africa, October 1996
- 52nd WMA General Assembly, Edinburgh, Scotland, October 2000

I. Basic Principles

1. Biomedical research involving human subjects must conform to generally accepted scientific principles and should be based on adequately performed laboratory and animal experimentation and on a thorough knowledge of the scientific literature.

2. The design and performance of each experimental procedure involving human subjects should be clearly formulated in an experimental protocol which should be transmitted for consideration, comment and guidance to a specially appointed committee independent of the investigator and the sponsor provided that this independent committee is in conformity with the laws and regulations of the country in which the research experiment is performed.

3. Biomedical research involving human subjects should be conducted only by scientifically qualified persons and under the supervision of a clinically competent medical person. The responsibility for the human subject must always rest with a medically qualified person and never rest on the subject of the research, even though the subject has given his or her consent.

4. Biomedical research involving human subjects cannot legitimately be carried out unless the importance of the objective is in proportion to the inherent risk to the subject.

5. Every biomedical research project involving human subjects should be preceded by careful assessment of predictable risks in comparison with foreseeable benefits to the subject or to others. Concern for the interests of the subject must always prevail over the interests of science and society.

6. The right of the research subject to safeguard his or her integrity must always be respected. Every precaution should be taken to respect the privacy of the subject and to minimize the impact of the study on the subject's physical and mental integrity and on the personality of the subject.

7. Physicians should abstain from engaging in research projects involving human subjects unless they are satisfied that the hazards involved are believed to be predictable. Physicians should cease any investigation if the hazards are found to outweigh the potential benefits.

8. In publication of the results of his or her research, the physician is obliged to preserve the accuracy of the results. Reports of experimentation not in accordance with the principles laid down in this Declaration should not be accepted for publication.

9. In any research on human beings, each potential subject must be adequately informed of the aims, methods, anticipated benefits and potential hazards of the study and the discomfort it may entail. He or she should be informed that he or she is at liberty to abstain from participation in the study and that he or she is free to withdraw his or her consent to participation at any time. The physician should then obtain the subject's freely-given informed consent, preferably in writing.

10. When obtaining informed consent for the research project the physician should be particularly cautious if the subject is in a dependent relationship to him or her or may consent under duress. In that case the informed consent should be obtained by a physician who Is not engaged in the investigation and who is completely independent of this official relationship.

11. In case of legal incompetence, informed consent should be obtained from the legal guardian in accordance with national legislation. Where physical or mental incapacity makes it impossible to obtain informed consent, or when the subject is a minor, permission from the responsible relative replaces that of the subject in accordance with national legislation.

    Whenever the minor child is in fact able to give a consent, the minor's consent must be obtained in addition to the consent of the minor's legal guardian.

12. The research protocol should always contain a statement of the ethical considerations involved and should indicate that the principles enunciated in the present Declaration are complied with.

Glossary

Achondroplasia A type of autosomal dominant genetic disorder that is a common cause of short stature, commonly referred to as dwarfism. Average adult height is 4'3" for males and 4' for females. It is characterized by abnormal bone growth that results in short stature with disproportionately short arms and legs, a large head, and characteristic facial features. Molecular genetic testing can be used to detect a mutation in the *FGFR3* gene. Such testing detects mutations in 99 percent of affected individuals.

Anonymized Data that cannot be traced back to their donor.

Antisense nucleotides Antisense nucleotides are strings of RNA or DNA that are complementary to "sense" strands of nucleotides. They bind to and inactivate these sense strands. They have been used in research and may become useful for therapy of certain diseases.

Biomarker Biomarkers are proteins, gene variants, or other biochemicals that indicate or can measure a biological process. Detecting biomarkers specific to disease can aid in the identification of risk, diagnosis, or treatment of affected individuals who do not exhibit symptoms.

Computational biology Bioinformatics or computational biology is the use of techniques from applied mathematics, informatics, statistics, and computer science to solve biological problems. Research in computational biology often overlaps with systems biology. Major research efforts in the field include sequence alignment, gene finding, genome assembly, protein structure alignment, protein structure prediction, prediction of gene expression and protein-protein interactions, and the modeling of evolution.

Dihybrid A dihybrid cross involves a study of inheritance patterns for organisms differing in two traits. Mendel invented the dihybrid cross to determine whether different traits of pea plants, such as flower color and seed shape, were inherited independently.

Disomy Refers to a normal chromosome pair as in the inheritance of both copies of a chromosome from the same parent.

Episomes A genetic element in bacteria that can replicate freely in the cytoplasm, or be inserted into the main bacterial chromosome and replicate with the chromosome.

Fragile X Fragile X syndrome is the most common inherited cause of mental impairment, and the most common known cause of autism. Fragile X syndrome is a genetic disorder caused by a mutation of the FMR1 gene on the X chromosome, a mutation found in 1 out of every 2,000 males and 1 out of every 4,000 females. Typically the FMR1 gene contains between 6 and 53 repeats of the CGG codon. In people with the disorder, the FMR1 allele has more than 230 repeats.

Genomics The study of genes and their functions.

Genotype The entire genetic identity of an individual, including alleles, or gene forms, that do not show as outward characteristics.

Germ-line Inherited material that comes from the eggs or sperm and is passed on to offspring

Germ-line gene therapy An evolving technique used to treat inherited diseases. The medical procedure involves either replacing, manipulating, or supplementing nonfunctional genes with healthy genes, and notably would affect the germ-line.

Haplotypes The genetic constitution of individuals with respect to one member of a pair of genes; sets of single alleles or closely linked genes that tend to be inherited together, such as those of the major histocompatibility complex; portions of phenotypes determined by genes located on one of a pair of chromosomes.

Hepatocytes These liver cells make up 70 to 80 percent of the cytoplasmic mass of the liver. These cells are involved in protein synthesis; protein storage and transformation of carbohydrates; synthesis of cholesterol, bile salts, and phospholipids; and detoxification, modification, and excretion of exogenous and endogenous substances. The hepatocyte also initiates the formation and secretion of bile.

HIPAA HIPAA is the United States Health Insurance Portability and Accountability Act of 1996. There are two sections to the Act. HIPAA Title I deals with protecting health insurance coverage for people who lose or change jobs. HIPAA Title II includes an administrative simplification section that deals with the standardization of health care–related information systems. HIPAA seeks to establish standardized mechanisms for electronic data interchange (EDI), security, and confidentiality of all health care–related data. The Act mandatesstandardized formats for all patient health, administrative, and financial data; unique identifiers (ID numbers) for each health care entity, including individuals, employers, health plans, and health care providers; and security mechanisms to ensure confidentiality and data integrity for any information that identifies an individual.

Karyotype A karyotype is an organized profile of a person's chromosomes. In a karyotype, chromosomes are arranged and numbered by size, from largest to smallest. A picture of the chromosomes in a cell that is used to check for abnormalities. A karyotype is created by staining the chromosomes with dye and photographing them through a microscope. The photograph is then cut up and rearranged so that the chromosomes are lined up into corresponding pairs.

Molecular genetic testing Analyzing DNA to look for a genetic alteration that may indicate an increased risk for developing a specific disease or disorder. Synonymous terms are DNA base testing, genetic testing, or genetic screening.

Monohybrid A monohybrid cross is a cross between individuals who are identically heterozygous at one locus, for example, Bb x Bb (see the Punnett square below). Monohybrid inheritance is the inheritance of a single characteristic. The different forms of the characteristic are usually controlled by different alleles of the same gene. For example, a monohybrid cross between two pure-breeding plants (homozygous for their respective traits), one with yellow seeds (the dominant trait) and one with green seeds (the recessive trait), would be expected to produce an F1 (first) generation with only yellow seeds because the allele for yellow seeds is dominant to that of green. A monohybrid cross compares only one trait.

Multipotent Multipotent and pluripotent cells are both referred to as "stem cells." Pluripotent stem cells are the precursors to multipotent stem cells and eventually develop into nearly all the cells required by humans, scientists say they have great medical-treatment applications. Through various experiments with animals, scientists have shown they can direct the specialization of pluripotent stem cells. In other words, the cells can be manipulated into virtually any specialized cell.

Myocytes A muscle fiber (American usage) or muscle fibre (British usage) is a single cell of a muscle. Muscle fibers contain many myofibrils, the contractile unit of muscles.

Nutrigenomics Nutrigenomics is the application of the sciences of genomics, transcriptomics, proteomics, and metabolomics to human nutrition, especially the relationship between nutrition and health. Nutrition and health research is focused on the prevention of disease by optimizing and maintaining cellular, tissue, organ, and whole-body homeostasis. This requires understanding, and ultimately regulating, a multitude of nutrient-related interactions at the gene, protein, and metabolic levels.

Oocyte cytoplasm Refers to a cell from which an egg or ovum develops by meiosis; a female gametocyte. Oocyte cytoplasm has the capacity to dedifferentiate somatic cells during maturation and has been investigated for its ability to facilitate genetic selection by oocyte transfer to prevent disease.

Phenotypic expression The phenotype of an individual organism describes one of its traits or characteristics that is measurable and that is expressed in only a subset of the individuals within that population Examples include "blue eyes" or "aggressive behavior." Some phenotypes are controlled entirely by the individual's genes. Others are controlled by genes but are significantly affected by extragenetic or environmental factors.

Pre-implantation diagnosis A procedure used to decrease the chance of a particular genetic condition for which the fetus is specifically at risk by testing one cell removed from early embryos conceived by in vitro fertilization and transferring to the mother's uterus only those embryos determined not to have inherited the mutation in question.

Proteomics The study of proteins and their functions.

Sensitivity Test sensitivity is the ability of a test to detect individuals with a disease. It is a posteriori analysis that compares the results of a test with a gold standard. Sensitivity is meant to ascertain test performance and not really to diagnose a disease. If a test is used to make a diagnosis, then a test with high sensitivity is used to rule out diseases.

Sickle cell disease Sickle cell disease is an inherited blood disorder that affects red blood cells. People with sickle cell disease have red blood cells that contain mostly hemoglobin* S, an abnormal type of hemoglobin. People with sickle cell conditions make a different form of hemoglobin A, called hemoglobin S (S stands for sickle). There are several types of sickle cell disease. The most common are sickle cell anemia (SS), sickle-hemoglobin C disease (SC) sickle beta-plus thalassemia, and sickle beta-zero thalassemia.

SNPs (single nucleotide polymorphisms) Specific positions in a DNA sequence that differ at one base. For example, the DNA sequences CCTGAA and CCCGAA differ in the third position—one sequence has the base T and the other the base C.

Specificity Test specificity refers to the ability of the test to identify specifically that which it is designed to detect and not other phenomenon.

Social construct   The meaning of a concept is defined by society not empirically determined facts. For example, consider that until relatively recently, rape within marriage was considered normal sexual activity and not criminal behavior.

Somatic cells   A somatic cell is generally taken to mean any cell forming the body of an organism. The word "somatic" is derived from the Greek word *sōma* (σῶμα), meaning "body." Somatic cells, by definition, are not germ-line cells. In mammals, germ-line cells are the spermatozoa and ova (also known as gametes), which fuse during fertilization to produce a cell called a zygote, from which the entire mammalian embryo develops. Every other cell type in the mammalian body—apart from the sperm and ova, the cells from which they are made (gametocytes) and undifferentiated stem cells—is a somatic cell. Internal organs, skin, bones, blood, and connective tissue are all made up of somatic cells.

Teratogenic   Meaning "to cause birth defects," teratology as a medical term was popularized in the 1960s by Dr. David W. Smith of the University of Washington Medical School, one of the researchers who became known in 1973 for the discovery of Fetal alcohol syndrome. With greater understanding of the origins of birth defects, the field of teratology now overlaps with other fields of medicine, including developmental biology, embryology, and genetics.

Totipotent   Cells that have the ability to develop into any of the many different cell types that make up multicellular organisms. Embryos are composed of large numbers of totipotent cells, which decline in number as development proceeds and cell specialization begins to occur. Adults have a much more limited ability to produce totipotent cells than do embryos.

Toxicogenomics   A new scientific subdiscipline that combines the emerging technologies of genomics and bioinformatics to identify and characterize mechanisms of action of known and suspected toxicants. Currently, the premier toxicogenomic tools are the DNA microarray and the DNA chip, which are used for the simultaneous monitoring of expression levels of hundreds to thousands of genes.

Xeno-transplantation   The transplantation of tissue from one species is tissue of another species.

Selected Bibliography

Ad Hoc Committee on Genetic Testing/Insurance Issues. *American Journal of Human Genetics* 56, no. 327 (1995): 327–31.

American Civil Liberties Union. "Genetic Discrimination in the Workplace Fact Sheet." Available at http://www.aclu.org.

Andrews, L., Fullarton, J. E., Holtzman, N. A., and Motulsky, A. G. *Assessing Genetic Risks: Implications for Health and Social Policy.* Washington, DC: National Academy Press, 1994.

Annas, G. "Privacy Rules for DNA Databanks: Protecting Coded 'Future Diaries'." *Journal of the American Medical Association* 270, no. 19 (November 17, 1993): 2346–50.

Baird, P. "Identifying People's Genes: Ethical Aspects of DNA Sampling in Populations." *Perspectives in Biology and Medicine* 38, no. 2 (Winter 1995): 159–66.

Baker, C. *Behavioral Genetics.* Washington, DC: American Association for the Advancement of Science, 2004.

"Boy Bullies' Bad Genes to Blame." Available at http://www.abc.netau/science/newsstories/19851.html.

Brambati, B., and Tului, L. "Curr Opin." *Obstetrics and Gynecology* 17 (2005): 197–201.

Bray, G. A. "Epidemiology, Trends, and Morbidities of Obesity and the Metabolic Syndrome." *Endocrine* 29, no. 1 (February 2006): 109–18.

Brennan, P. A., Mednick, S. A., and Jacobsen, B. "Assessing the Role of Genetics in Crime Using Adoption Cohorts." *Ciba Foundation Symposium* 194 (1996): 115–23, discussion 123–28.

Campbell, K. H. S., McWhir, J., Ritchie, W. A., and Wilmut, I. "Sheep Cloned by Nuclear Transfer From a Cultured Cell Line." *Nature* 380 (1996): 64–66.

Center for Genetics and Society. "Failure to Pass Federal Cloning, 1997-2003." Available at http://www.genetics-and-society.org/policies/us/cloning.html#3a.

Chiappe, A., Cottini, S., Dayer, P., and Desmeules, J. "Brief Report: Codeine Intoxication Associated with Ultra Rapid CYP2D6 Metabolism." *New England Journal of Medicine* 351 (December 30, 2004): 2827–31.

Colin, C. "Reading Genes in Black and White." University of Florida professor claims he has evidence blacks are intellectually inferior, April 26, 1999. Available at http://www.salon.com/books/it/1999/04/26/genetics/print.html.

Collins, F., and Watson, J. "Genetic Discrimination: Time to Act." *Science* 2, no. 31 (October 2003).

Council for Responsible Genetics. Available at http://www.crg.org.

Cunningham, G. "The Genetics Revolution: Ethical, Legal, and Insurance Concerns." *Postgraduate Medicine* 108, no. 1 (2000): 193–202.

Cunningham, G., Deftos, L. J., Nowlan, W. J., and Clayton, E. W. "Correspondence." *New England Journal of Medicine* 349 (November 6, 2003): 1870–72.

Davenport, C. B. *Heredity in Relation to Eugenics.* New York: Holt, 1911.

Davis, S., and Reynolds, L. A. "Genetic Discrimination: Why Should We Care?" In "Genetic Issues in Mental Retardation." *The Arc* 1, no. 2 (1966). Available at http://www.thearc.org/pdf/gbr02.pdf.

de Lorgeril, M., Renaud, S., Mamelle, N., Salen, P., Martin, J. L., Monjoud, I., Guidollet, J., Touboul, P., and Delaye, J. "Mediterranean Alpha-Linolenic Acid-Rich Diet in Secondary Prevention of Coronary Heart Disease." *Lancet* 343 (1994): 1454–59.

de Lorgeril, M., Salen, P., Martin, J. L., Mamelle, N., Monjoud, I., Touboul, P., and Delaye, J. "Effect of a Mediterranean Type of Diet on the Rate of Cardiovascular Complications in Patients with Coronary Artery Disease." *Journal of the American College of Cardiology* 28 (1996): 1103–8.

Dhanda, R. K. G. *Guiding Icarus: Merging Bioethics with Corporate Interests.* New York: Wiley, 2002.

Diver, C., and Cohen, J. "Genophobia: What Is Wrong with Genetic Discrimination?" *University of Pennsylvania Law Review* 149, no. 5 (May 2001): 1439–82.

Food and Agriculture Organization of the United Nations. "FAO Corporate Document Repository." Available at http://www.fao.org/DOCREP/003/X9602E/x9602e04.htm#P0_0.

GeneForum. Available at http://www.geneforum.org.

Genetic Information and the Workplace. Available at http://www.eeoc.gov.

Genetic Privacy Act. Available at http://www.ornl.gov/sci/techresources/Human_Genome/resource/privacy/privacy1.html.

Hallman, W. K., and Hebden, W. C. "American Opinions of GM Food: Awareness, Knowledge, and Implications for Education." *Choices.* Available at http://www.choicesmagazine.org/2005-4/GMOs/2005-4-05.htm.

Hampton, T. "Expanded Newborn Genetic Testing Urged." *Journal of the American Medical Association* 292, no. 19 (November 17, 2004): 2325–26.

Hardin, G. "Lifeboat Ethics: The Case Against Helping the Poor." In *Morality in Practice,* ed. J. Sterba. Belmont, CA: Wadsworth, 1988.

Heuser, R. "From Secrecy to Openness: The U.S. Historical Perspective." Available at http://www.wits.ac.za/saha/publications/contrib_heusser.pdf.

Holtzman, N. A. "Expanding Newborn Screening: How Good Is the Evidence?" (Editorial). *Journal of the American Medical Association* 290, no. 19 (November 19, 2003): 2606.

Hood, E. "Pharmacogenetics: Targeting Medicine's Future—Environmental Health Perspectives." *Journal of the National Institute of Environmental Health Sciences* 111, no. 11 (August 2003).

Human Genome Project. *The Human Genome Project: Science, Law, and Social Change in the 21st Century.* Whitehead Policy Symposium. Cambridge, MA: Whitehead Institute for Biomedical Research and the American Society for Law, Medicine & Ethics, 1998. (Conference on computer disk.) For genetic splicing/recombinant DNA controversy, see http://www.nih.gov/news/NIH-Record/01_22_2002/story02.htm and http://www.accessexcellence.org/RC/AB/IE/wholesome.html.

Iowa State University. Biosafety Institute for Genetically Modified Agricultural Products. "GAO: Testing of GM Foods Is Adequate, Monitoring of Health Risks Not Needed," May 2002. Available at http://www.bigmap.iastate.edu/www/285.htm.

Jones, J. H. *Bad Blood: The Tuskegee Syphilis Experiment,* rev. ed. New York: Free Press, 1993.

Jonsen, A. R., Siegler, M., and Winslade, W. J. *Clinical Ethics,* 4th ed. New York: McGraw-Hill, 1998.

Khoury, M., and Burke, W. *Genetics and Public Health in the 21st Century: Using Genetic Information to Improve Health and Prevent Disease.* Oxford: Oxford University Press, 2000.

Kneller, R. "Genetic Privacy and Discrimination." In *Bioethics and the Impact of Human Genome Research in the 21st Century.* Tokyo: Eubios Ethics Institute, Tokyo University, 2001.

Knoppers, B. M., ed. *Populations and Genetics: Legal and Socio-ethical Perspectives.* Leiden: Martinus Nijhoff, 2003.

Lapham, E. V., Kozma, C., and Weiss, J. O. "Genetic Discrimination: Perspectives of Consumers." *Science* 274 (October 25, 1996): 621–23.

Liu, H. C., He, Z. Y., Mele, C. A., Veeck, L. L., Davis, O., and Rosenwaks, Z. H. "Endometrial Stromal Cells Improve Embryo Quality by Enhancing the Expression of Insulin-Like Growth Factors and Their Receptors in Cocultured Human Preimplantation Embryos." *Fertility and Sterility* 71 (1999): 361–67.

Long Bets Foundation. "Predictions." Available at http://www.longbets.org/predictions.

Lyons, M. J. "A Twin Study of Self-Reported Criminal Behavior." *Ciba Foundation Symposium* 194 (1996): 61–70, discussion 70–75.

"Male Aggression? It's Only Natural." Available at http://www.abc.net.au/science/news/health/HealthRepublish_394779.htm.

Mange, E. J., and Mange, A. P. *Genetics: Human Aspects*, 2nd ed. Sunderland, MA: Sinauer Associates, 1990.

Markel, H. "Becoming a Physician: 'I Swear by Apollo'—on Taking the Hippocratic Oath." *New England Journal of Medicine* 350 (2004): 2026–29.

McGuffin, P., Riley, B., and Plomin, R. "Genomics and Behavior: Toward Behavioral Genomics." *Science* 291, no. 5507 (February 16, 2001): 1232–49.

Mitka, M. "Neonatal Screening Varies by State of Birth." *Journal of the American Medical Association* 284, no. 116 (October 25, 2000): 2044.

Murray, T. "Genetics and the Moral Mission of Health Insurance." *Hastings Center Report* 22, no. 6 (1992): 12–17.

National Association of Health Underwriters. "Genetic Discrimination." Available at http://www.nahu.org.

National Conference of State Legislatures. State Privacy Laws. Available at http://www.ncsl.org/programs/health/genetics/prt.htm.

National Institutes of Health. "Pharmacogenetics Research Network." Available at http://www.nigms.nih.gov/pharmacogenetics.

"New Tools for Personalized Medicine." *Wall Street Journal*, January 11, 2005.

President's Council on Bioethics. Available at http://www.bioethics.gov.

Renaud, S., de Lorgeril, M., Delaye, J., Guidollet, J., Jacquard, F., Mamelle, N., Martin, J. L., Monjoud, I., Guidollet, J., and Touboul, P. "Cretan Mediterranean Diet for Prevention of Coronary Heart Disease." *American Journal of Clinical Nutrition* 61, suppl. 6 (1995): 1360S–67S.

Rothenberg, K., Fuller, B., Rothstein, M., Duster, T., Ellis Kahn, M. J., Cunningham, R., Fine, B., Hudson, K., King, M. C., Murphy, P., Swergold, G., and Collins, F. "Genetic Information and the Workplace: Legislative Approaches and Policy Challenges." *Science* 275 (March 21, 1997).

Schatz, G. "Health Records Privacy and Confidentiality: Pending Questions." *Journal of Contemporary Health Law and Policy* 18 (2002): 685–91.

Schnieke, A. E., Kind, A. J., Richie, W. A., Mycock, K., Scott, A. R., Ritchie, M., Wilmut, I., Colman, A., and Campbell, K. H. "Human Factor XI Transgenic Sheep Produced by Transfer of Nuclei From Transfected Fetal Fibroblasts." *Science* 278 (1997): 2130–33.

Service, R. "Surviving the Blockbuster Syndrome." *Science* 303 (March 19, 2004): 1796–99.

Simpolous, A., and Ordorvas, J. eds. *Nutrigenetics and Nutrigenomics*. World Review of Nutrition and Dietetics, vol. 93. Basel: Karger, 2004.

Simopoulos, A. P., and Robinson, J. *The Omega Diet: The Lifesaving Nutritional Program Based on the Diet of the Island of Crete.* San Francisco: HarperPerennial, 1999.

Tenner, E. *Why Things Bite Back: Technology and the Revenge of Unintended Consequences.* New York: Vintage Books, 1997.

Thompson, M., McInnes, R., and Willard, H. *Genetics in Medicine,* 5th ed. New York: W. B. Saunders, 1991.

United Network for Organ Sharing. Available at http://www.unos.org.

U.S. Census Bureau. "Total Midyear Population for the World: 1950–2050." Available athttp://www.census.gov/ipc/www/worldpop.html.

U.S. Congress Office of Technology Assessment. "Development of the Right to Privacy in Information." Protecting Privacy in Computerized Medical Information, OTA-TCT-576, September 1993.

U.S. Department of Energy. Genome Program. Available at http://www.ornl. gov/sci/techresources/Human_Genome/elsi/elsi.shtml.

Webber, W. *Pharmacogenetics.* New York: Oxford University Press, 1997.

Weindling, P. J. *Nazi Medicine and the Nuremberg Trials: From Medical War Crimes to Informed Consent.* London: Palgrave Macmillan, 2005.

"Whose Medical History Is It Anyway?" *USA Today,* April 8, 2001. Available at http://www.usatoday.com/news/health/2001-04-08-privacy.htm. See also http://www.matr.gen.vcu.edu/guides.htm.

Wilmut, I., Schnieke, A. E., McWhir, J., Kind, A. J., and Campbell, K. H. S. "Viable Offspring Derived from Fetal and Adult Mammalian Cells." *Nature* 385 (1997): 810–13.

Wingrove, K., Norris, J., Barton, P., and Hagerman, R. "Experiences and Attitudes Concerning Genetic Testing and Insurance in a Colorado Population: A Survey of Families Diagnosed with Fragile X Syndrome." *American Journal of Medical Genetics* 64 (1996): 378–81.

World Health Organization. The Declaration of Helsinki: Recommendations Guiding Medical Doctors in Biomedical Research Involving Human Subjects, 1964.

World Hunger Year. "Domestic Hunger and Policy Facts." Available at http://www.worldhungeryear.org/info_center/just_facts.asp.

Wright Clayton, E. "Ethical, Legal, and Social Implications of Genomic Medicine." *New England Journal of Medicine* 349, no. 6 (August 7, 2003): 562–69.

Yesley, M. *Ethical, Legal, and Social Implications of the Human Genome Project.* U.S. Department of Energy, Office of Energy Research. Springfield, VA: National Technical Information Service, 1993.

Youth for International Socialism. Interesting statistics. Available at http://www.newyouth.com/interestingstats.asp#Malnourishment%20and%20Starvatio.

Index

About the Author

CAROL ISAACSON BARASH is Principal of Genetics, Ethics & Policy Consulting, which she founded in 1994. She has worked with leading public and private institutions around the world to optimize the ethical integration of new genomic technologies and ensure that they comply with the highest ethical standards. Her work has been featured in *Newsweek, BioIT World, BioNews,* MSNBC, and other news outlets. She has published more than 25 articles and is contributing author to several books, including *The Double-Edged Helix, Genetic Connections, Risk vs. Risk: Tradeoffs in Protecting Health and Environment,* and others. Previously, she directed the first federally funded study of genetic discrimination and has taught ethics courses at Boston University, Bentley College, and Lowell University.